Electrodeposition of Chromium from Chromic Acid Solutions

Pergamon Books of Related Interest

Coulson & Richardson – Chemical Engineering

Gabe – Principles of Metal Surface Treatment and Protection

Santholzer – Five Language Dictionary of Surface Coatings, Plating, Products Finishing, Corrosion, Plastics and Rubber

Sobecka, et al. – Dictionary of Chemistry and Chemical Technology – 2nd Edition

Weisser & Landa – Sulphide Catalysts

Electrodeposition of Chromium from Chromic Acid Solutions

George Dubpernell
M&T Chemicals Inc.
Ferndale, Michigan 48220

PERGAMON PRESS, INC.
New York/Oxford /Toronto/Sydney/Frankfurt/Paris

Pergamon Press Offices:

U.S.A.	Pergamon Press Inc., Maxwell House, Fairview Park, Elmsford, New York 10523, U.S.A.
U.K.	Pergamon Press Ltd., Headington Hill Hall, Oxford OX3, OBW, England
CANADA	Pergamon of Canada, Ltd., 207 Queen's Quay West, Toronto 1, Canada
AUSTRALIA	Pergamon Press (Aust) Pty. Ltd., 19a Boundary Street, Rushcutters Bay, N.S.W. 2011, Australia
FRANCE	Pergamon Press SARL, 24 rue des Ecoles, 75240 Paris, Cedex 05, France
WEST GERMANY	Pergamon Press GmbH, 6242 Kronberg/Taunus, Frankfurt-am-Main, West Germany

Library of Congress Cataloging in Publication Data
Dubpernell, George, 1901 -
Electrodeposition of chromium from chromic acid solutions.

1. Chromium-plating. I. Title.
TS692.C4D8 1977 671.7'3 77-549
ISBN 0-08-021925-X

Printed in the United States of America

Contents

Preface

It is with pleasure that a book is offered on the behavior of the major catalysts used in the electrodeposition of chromium from chromic acid solutions, to commemorate 50 years of commercial chromium plating following the introduction of the process by Colin G. Fink and Charles H. Eldridge in 1924-1926. Even after 50 years the behavior of these catalysts is little understood in the industry, and their use is largely empirical. It is hoped that the present book will be helpful to all concerned, and encourage further development.

The story of the introduction of chromium deposition by Fink and Eldridge was told in Plating 47, 35-53, 1960, and their biographies appear in "American Chemists and Chemical Engineers," edited by Wyndham D. Miles and published by the American Chemical Society, 544 pages, 1976.

Numerous articles, patents, and research reports have been searched with great care to collect as much quantitative information as possible on the three major catalysts which control the process – sulfate, fluoride and chloride. This quantitative information is concentrated in the current efficiency-catalyst curves

(Figs. 5, 6, 7, 21, 22, 23, 24 and 26) supplemented by other current efficiency curves (Figs. 4, 17, 19 and 25). The current efficiency is of prime importance, since it may vary from zero to 30 or 40 percent commonly, and is a direct index of how much chromium is being obtained, as well as a clue to its quality and usefulness.

The industrial usefulness of chromium deposits is bound up with their very great adhesion to the basis metal when they are applied from warm solutions. The basis metal surface is usually in a damaged condition due to having been ground, machined, or etched previously, and chromium plate may be said to have considerable "healing" ability, in addition to providing hardness, wear resistance and other desirable surface qualities. Thus, it is often called a "putting on tool" for bringing parts up to size when necessary.

Introductory chapters on history, economic importance and the two common forms of deposited chromium are provided. The catalyst balance or its ratio to chromic acid concentration is discussed to show the importance of this controlling factor. The review of empirical tests for catalyst concentration and bath balance should serve to further emphasize the control exerted by the catalyst concentration, and to give the experienced chemist greater insight into the nature of the process. The final chapter on fluoride catalysts it is hoped will bring a measure of order and understanding into what is indeed a complicated subject.

The book should be valuable to executives and plant superintendents, and particularly to engineers, metallurgists and control chemists confronted with the need to interpret the results of chemical analyses and tests of baths used for chromium deposition. It should also be of interest to research workers in both the industrial and academic laboratories. Much of the chemistry and electrochemistry of complex salts is involved.

I should like to express my gratitude to M&T Chemicals Inc. for permission to publish and for commemorating 50 years of chromium plating, as well as to many colleagues and friends whose help and encouragement made this book possible. These include particularly Dr. E.J. Seyb, Frank Passal, J.A. Williams and P.C. Burnham. Last but not least I should mention my wife Grace without whose constant assistance, stimulation and support the book could not have been written.

Professor Colin G. Fink in his laboratory at Columbia University about 1925.

1
Historical

Chromium was not known to the ancients, and credit for discovery of the metal is given to Nicolas-Louis Vauquelin in Paris in 1797-1798 [1]. Vauquelin analyzed a Siberian red lead (crocoite, $PbCrO_4$) and found it to contain an unknown acid in addition to lead. He precipitated the lead as lead chloride and, after evaporating the residue, reduced it with carbon at a high temperature and obtained a network of gray interlacing metallic needles. Because of the many colored compounds Fourcroy and Haüy suggested the name chromium for the new metal.

In just a few years chromite ($FeO \cdot Cr_2O_3$) was identified as a widely distributed and readily available source of chromium chemicals [2-4]. It appears that known reserves of chromite or chromium ore, particularly in South Africa and Southern Rhodesia, are sufficient to last for the foreseeable future.

Thus, chromium chemicals were something of a 40-year-old novelty when Antoine Becquerel speculated on the results to be obtained by electrolyzing their solutions in his book in 1843 [5]. He said that one would probably succeed in obtaining chromium using chromic chloride; one could also try the chromic sulfate soluble in water, or the green chromic nitrate which dissolves easily.

Probably the first to deposit chromium was Junot de Bussy in 1848-1849 [6]. Junot's patents were not very clear as he claimed to have found ways to deposit a number of metals which we still cannot deposit from aqueous solution. However, his purpose was quite clear. He wanted to break the monopoly of Ruolz and Elkington on cyanide gold and silver plating by depositing an undercoat of a corrosion resistant metal on the basis metal, and then plating with gold or silver with the corrosive chloride or other baths which were in the public domain, such as those of Becquerel or De la Rive. Thus, he claimed to deposit undercoats of silicon, chromium, vanadium, tungsten, titanium, molybdenum, tin and tantalum to replace platinum which was too expensive.

In his patent of addition in 1849, Junot disclosed that he used a saturated solution of chromium chloride filtered onto an excess of sodium chloride. The solution was then diluted to 10% with distilled water, boiled, and filtered again with great care, when it had a beautiful clear amethyst color by transmitted light. At the end of this addition patent, Junot commented that when his new deposits were made to some thickness they had a whiteness and brightness that made any subsequent deposit of silver unnecessary, and had the further advantages of resisting the action of hydrogen sulfide completely, and of having greater wear resistance.

In a later French patent in 1855, Junot merely reaffirmed that he used a chromium chloride made by dissolving hydrated chromic oxide in hydrochloric acid and evaporating to dryness to remove excess hydrochloric acid [7]. The British patent of 1852 is uninformative but indicates that he used about 10 g/l of chromium as the double chloride of sodium and ammonia [8].

At about this same time, in 1854, Professor Robert Bunsen in Heidelberg investigated the effect of current density on the electrodeposition of chromium from chromous chloride solutions containing chromic chloride, in a two compartment cell [9]. The cell was made with a carbon crucible as anode and containing some hydrochloric acid in the anode compartment. In the center a small porous ceramic container was placed and contained the solution and a small platinum strip cathode. With this cell and concentrated, boiling hot solutions, Bunsen easily obtained small sheets of metallic chromium more than half a square centimeter in area. They were quite brittle; the surface next to the platinum cathode was smooth and bright.

The chromium looked like iron, but was more resistant to damp air. It dissolved with difficulty in hydrochloric acid and dilute sulfuric acid to form chromous salts. Nitric acid, even boiling, scarcely attacked the metal. The specific gravity was about one-seventh higher than previously given in the textbooks, that is, about 6.9 instead of 5.9.

In the second edition of his text book of electrochemistry in 1864 [10], Becquerel's only comments on chromium deposition were that chromium salts were difficult to reduce to the metal with electricity, and to give a description of Bunsen's work.

In 1856 Anton Geuther in Goettingen, Germany discovered the deposition of chromium from chromic acid solutions while studying elements with more than one valence and trying to disprove Faraday's Law [11]. His work and the detailed history of chromium deposition from chromic acid baths are reviewed elsewhere [12].

Thus it was that more than a century ago some of the main features of the electrodeposition of chromium were anticipated, even though the difficulties of accomplishing this operation postponed the introduction of commercial chromium plating from chromic acid solutions until about 1925, and we are still trying to overcome the difficulties of depositing the metal from trivalent baths. The desirable properties of the metal became known by the 1850s, and those concerned with the electrodeposition of metals generally experimented with chromium. Thus, Isaac Adams Jr., who introduced commercial nickel plating in 1869 [13], frequently discussed the possibility of chromium plating in his correspondence in this period.

The basic difficulty of electrodepositing metallic chromium is evident from Fig. 1, a periodic chart published by the writer in 1946 [14] and brought up to date in 1975 [15]. It can be noted that chromium is not included in the group of metals easily electrodeposited at high current efficiencies from aqueous solutions. Instead, chromium is grouped with metals only reduced from their solutions with difficulty at high cathode potentials, which tend to the evolution of hydrogen simultaneously and give reduction products in the solution such as trivalent chromium in chromic acid solutions or bivalent chromium in trivalent solutions.

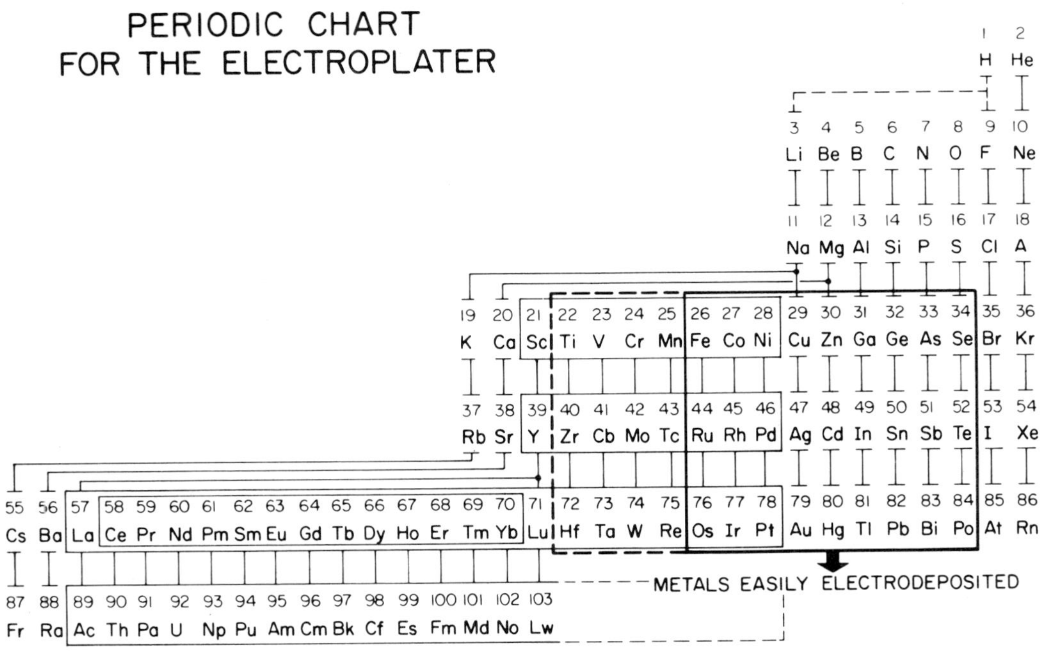

Fig. 1. Dubpernell periodic chart adapted to emphasize the ease or difficulty of depositing each element.

References

1. Weeks, M.E. "Discovery of the Elements," 6th edition, 1956, pp. 270-283, Easton, Pa.: Journal of Chemical Education.
2. Udy, M.J. (Ed.. "Chromium," 1956, 2 volumes, New York: Reinhold Publishing Corporation.
3. Sully, A.H., and Brandes, E.A. "Chromium," 2nd edition, 1967, London: Butterworth & Company, Ltd.
4. Gmelins Handbuch der anorganische Chemie (Gmelin Handbook of Inorganic Chemistry), System No. 52, "Chromium," Weinheim/Bergstrasse: Verlag Chemie, GmbH.
   Part A, Section 1: History, Occurrence, Technology, 418 pages, 1962.
   Part A, Section 2: Electrochemistry, Chemical Reactions, Alloys, 312 pages, 1963.
   Part B: The Compounds, 942 pages, 1962.
   Part C: Coordination Compounds, 431 pages, 1965.
5. Becquerel, A.C. "Elements D'Electrochimie," 1843, Paris: Librarie de Firmin Didot Freres.
6. Junot de Bussy, French 3564 – November 17, 1848; certificate of addition, November 12, 1849.
7. Junot, Chaudron, French 13,902 – October 8, 1855.
8. Junot, C.J.E., British 1183 – December 28, 1852.

9. Bunsen, R., Pogg, Ann. **91**, 619-625, 1854; The Chemist, 1854, No. 11, pp. 685-686.
10. Becquerel, A.C. "Elements D'Electrochimie," 2nd edition, 1864, Paris: Librairie de Firmin Didot Freres, Fils et Cie.
11. Geuther, A., Liebig's Ann. **99**, 314-333, 1856.
12. Dubpernell, G., Plating **47**, 35-53, 1960.
13. Dubpernell, G., Plating **46**, 599-616, 1959.
14. Dubpernell, G., Proc. Am. Electroplaters' Soc. **33**, 244-257, 1946; Extracts: Iron Age **158**, 51, November 28, 1964; Metalloberflaeche, **1**, B116, 1949.
15. Dubpernell, G., Plating **62**, 327-334, 1975.

2
Economic Importance of Chromium Deposition

It is difficult to describe the economic importance of chromium plating in modern society, as its presence and functions are more often than not invisible or intangible. Of course, almost everyone is aware of decorative chromium plate and its ability to maintain bright metal surfaces in an untarnished condition year after year on automobiles, appliances, plumbing fixtures, and so on. However, how many passengers on a railroad train are aware that the long life and durability of the cylinders of the diesel engines drawing the train depend upon a lining of a special porous chromium plate? Or again, how many passengers on a modern jet plane stop to admire the hard chromium plate on the hydraulic pistons of the landing gear, or people watching heavy earth moving equipment note the gleaming chromium on its hydraulic rams?

The applications of chromium on modern machine parts are literally endless because of its great wear resistance, low coefficient of friction and non-wetting or non-galling properties, heat resistance and chemical resistance. Another important characteristic is the relative ease with which valuable or expensive worn parts may be salvaged and put back into use for as long or longer than the original period of service.

The story was told of how rapidly the engines of Air Force planes wore out under exposure to abrasive dust in North Africa during World War II. Replacements were not available under wartime conditions so worn cylinders were salvaged by porous chromium plating with channel type porosity, and were found to give much longer service than the original cylinders. This is confirmed by Diggin [1] who also records that porous chromium plated piston rings likewise enabled planes to stay in service at least six times longer without an overhaul.

In still another area, an extremely thin chromium plate underneath lacquer is fast replacing tin on the tinplate used for making tin cans, as it has good corrosion resistance and is less expensive [2]. This results in a saving of scarce tin for other purposes, and may make the empty cans more recyclicable since they are not contaminated with tin which is injurious in ferrous metallurgy. If one buys a "tin can" of food or beverage now, it may be made from tin free steel (TFS) and not be coated with tin at all.

The chromic acid used in chromium plating is rather readily produced from the iron-chromium ore, chromite, by means of an oxidizing roast of the pulverized ore [3-7]. The ore contains about 45%, Cr_2O_3 (31% Cr), and is a relatively inexpensive raw material selling for \$25.00 – \$27.00 per long ton [8]. This is equivalent to about \$0.012 per lb. of ore, or about \$0.039 per lb. of contained chromium. The sodium bichromate first produced from the ore sells for about \$0.18 per lb. or about \$0.52 per lb. of contained chromium [9]. Chromic acid produced from bichromate sells for \$0.36 per lb. or about \$0.70 per lb. of contained chromium [9]. In contrast, pure electrolytic or aluminothermic chromium sold for about \$1.00 per lb. in 1969, while a pound of chromium in the form of ferrochromium only cost from \$0.14 to \$0.28 per lb. according to the grade [10].

It is apparent from the above figures that chromic acid is a less expensive source of chromium than metallic chromium. Metallic chromium could be used for anodes for chromium plating except that it dissolves as chromic acid at about 100% current efficiency, whereas the plating of the metal from hexavalent solutions is generally only about one-seventh as efficient, or at most may be conducted at efficiencies of 20 to 25%.

The use of ferrochromium anodes would appear to offer economic possibilities, but limited investigations along this line to date have not been encouraging [11, 12]. A substantial amount of iron has to be disposed of, and there would probably be little possibility of reoxidizing the trivalent chromium formed at the cathode during chromium deposition. The latter operation would presumably have to be conducted separately. If the ferrochromium anodes dissolved well they would have the same disadvantage as chromium anodes, and the process would have to include the production of chromic acid or chromates as a byproduct. One goes around a circle of thinking in this regard, and in the end it seems to boil down to the electrolytic production of chromic acid in competition with that produced chemically, and then using it in a separate chromium plating operation with lead anodes as usual.

When commercial chromium plating was first introduced in 1924-1925, the consumption of chromic acid in the United States for other purposes (largely medicinal) was only a few thousand pounds per year [13]. The acid was rather promptly made available commercially with a purity higher than previous U.S. Pharmacopoeia requirements. This was one key to the immediate success of the process.

The process required pure chromic acid with an addition of 1% by weight of sulfate ion or sulfate radical, $SO_4^{=}$, or what is close to the same thing, about 1% of sulfuric acid, H_2SO_4. The new commercial chromic acid was consistently produced with a purity of less than 0.1% sulfate, although since World War II this is less strictly maintained and often amounts to several tenths of a percent.

The history of chromic acid production in the United States, and of its price variation, is given in Fig. 2 [5, 13, 14]. It is noteworthy that production and consumption still seem to be increasing, even though the higher price would tend to slow this down.

The exact percentage of chromic acid production used for chromium plating does not seem to be known with certainty, and is variously estimated from 67% upwards, based upon 75% consumed in metal finishing uses, and 90% of this for chromium plating [15]. A more recent figure is 90% for metal finishing purposes [16], and this would correspond to about 80 or 81% for chromium plating. The other metal finishing uses include chromate treatments of metals, anodizing of aluminum, and copper stripping. It thus appears that about 48,600,000 lbs. of chromic acid were used for chromium plating in the United States in 1973.

Chromium plating has a substantial advantage over the plating of other metals in that the supply of high grade ore, particularly in Rhodesia and South Africa, is so extensive, and the ore is readily converted to the required chemicals. Nickel, for example, must be extracted from ores containing only a few percent of the metal at best.

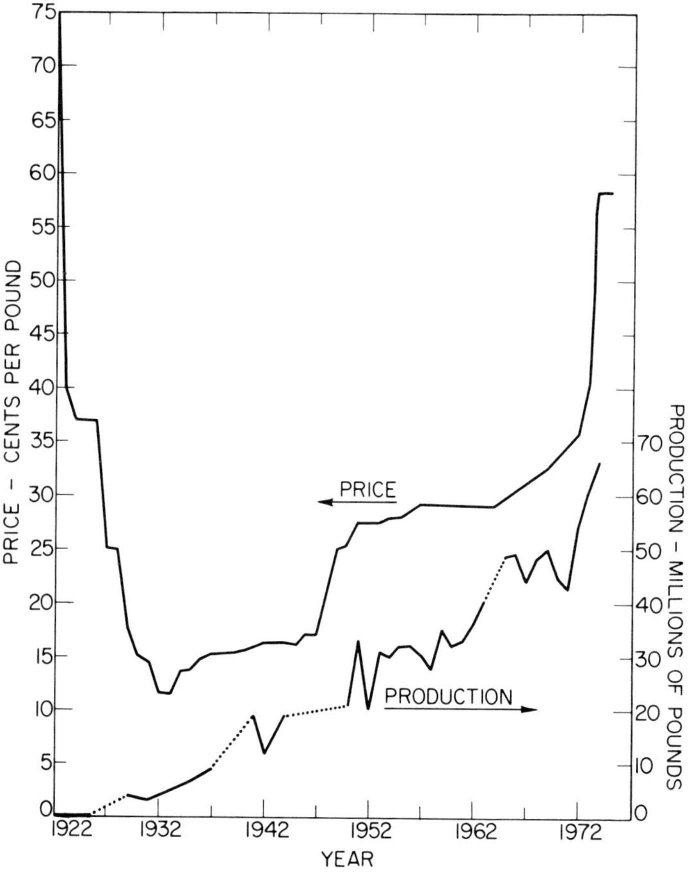

Fig. 2. Price and production of chromic acid in the United States.

This situation also promotes the use of pure chromium metal in stainless steel and other alloys. The high-chromium alloys such as Nichrome for heating elements require chromium metal of special purity. This was formerly supplied by the aluminothermic reduction of pure chromic oxide, but in recent years aluminothermic chromium has been replaced by electrolytic chromium to a great extent.

The Union Carbide Corporation, together with the U.S. Bureau of Mines, developed a process for the electrolytic production of chromium metal [17]. A plant was established in Marietta, Ohio which began commercial production in 1954. According to Kuhn [18] production at this plant is now about 3,000 long tons (6,720,000 lbs.) per year.

High-carbon ferrochromium is dissolved in sulfuric acid and anolyte solution from the two-compartment cell used. A pure ammonium chrome alum $[(NH_4)_2SO_4 \cdot Cr_2(SO_4)_3 \cdot 24H_2O]$ is produced which is dissolved and forms the cell feed to the cathode compartment of the electrolytic cell.

References

1. Diggin, M.B., Metal Finishing **45**, 78-80, June, 1947.
2. Smith, E.J., Blast Furnace and Steel Plant **55**, 801-808, 1967.
3. Combs, E.L., Plating **40**, 369-372, 377, 1953.
4. Copson, R.L. "Production of Chromium Chemicals," Chapter 8, pages 262-282 in Udy, M.J. (Ed.). "Chromium," Volume 1, 1956, New York: Reinhold Publishing Corporation.
5. Faith, W.L., Keyes, D.B., and Clark, R.L. "Industrial Chemicals," 3rd edition, 1965, New York: John Wiley & Sons, Inc. (Chapters on Chromic Acid and Sodium Bichromate, pages 272-275, and 685-689).
6. Vetter, J.J. "Chromium Compounds," in Kirk-Othmer Encyclopedia of Chemical Technology, Volume 3, pages 941-955, 1949.
7. W.H. Hartford and R.L. Copson. "Chromium Compounds," in Kirk-Othmer Encyclopedia of Chemical Technology, 2nd edition, Volume 5, pages 473-516, 1964.
8. U.S. Bureau of Mines, Mineral Industry Surveys, "Chromium in 1971," December 27, 1971.
9. Chemical Marketing Reporter, February 7, 1972.
10. Morning, J.L. "Chromium," pages 277-286 in 1969 U.S. Bureau of Mines Minerals Yearbook, Volume I-II.

11. Agladze, R.I., Soviet Electrochemistry (Proc. 4th Conference on Electrochemistry) Volume III, 30-34, 1961, New York: Consultants Bureau.
12. Agladze, R.I., (Ed.). "Gidroelektrometallurgiya Khroma-Sbornik Rabot" (Hydroelectrometallurgy of Chromium-Collected Works), 1959, Tbilisi: Akad. Nauk Gruzin, S.S.R. (Georgia).
13. W. Haynes. "American Chemical Industry," Volume IV, 1948, New York: D. Van Nostrand Company, Inc.
14. U.S. Department of Commerce, Bureau of the Census. "Current Industrial Reports; Inorganic Chemicals – 1970," Washington, D.C.: June, 1972, Series M 28A (70) – 14.
15. Chemical Week **75**, (4), 69-71, July 24, 1954.
16. Oil, Paint and Drug Reporter **195** (23), June 9, 1969, Chemical Marketing Reporter **202** (5), 9, 20-21, July 31, 1972.
17. Rosenbaum, J.B., Lloyd, R.R. and Merrill, C.C. "Electrowinning Chromium Metal," Bureau of Mines Report of Investigations 5322, 58 pages, March, 1957.
18. Kuhn, A.T. "Industrial Electrochemical Processes," page 212, London: Elsevier Publishing Company, 1971.

3
Two Common Forms of Deposited Chromium

It is not usually realized that chromium is quite commonly deposited from chromic acid solutions in two rather distinctively different forms having somewhat different properties and different X-ray structures. Some of the names used to designate these two forms of chromium are tabulated in Table 1.

Table 1

COLD SOLUTION FORM	WARM SOLUTION FORM
Cold chromium or dull chromium	Bright chromium
Hexagonal chromium	Cubic chromium
Hexagonal close packed (hcp)	Body centered cubic (bcc)
Chromium hydride	Cubic chromium
β-chromium	α-chromium

Cold chromium or the hexagonal hydride is the type of smooth, dull deposit obtained at temperatures of 20°C (68°F) and below. It is an unstable form of chromium and tends to revert to the stable cubic form in a few weeks at room temperature without changing its appearance. Deposits from 20°C to 40°C (68°F to 104°F) consist of mixtures of various proportions of hexagonal and cubic structure, while above 40°C (104°F) deposits are entirely cubic.

Attention is called to these two types of chromium plate at this point as an aid in reading and studying publications on the subject. Thus, many Russian articles in particular report investigations of chromium plating at 20°C (68°F) without mentioning that the deposits obtained are dull and bear little resemblance to the bright chromium plate obtained from warm solutions. The conclusions arrived at in investigations of cold baths may or may not apply to warm solutions. While in general there is a fairly close connection and a reasonable degree of applicability, one should always give some consideration to possible differences.

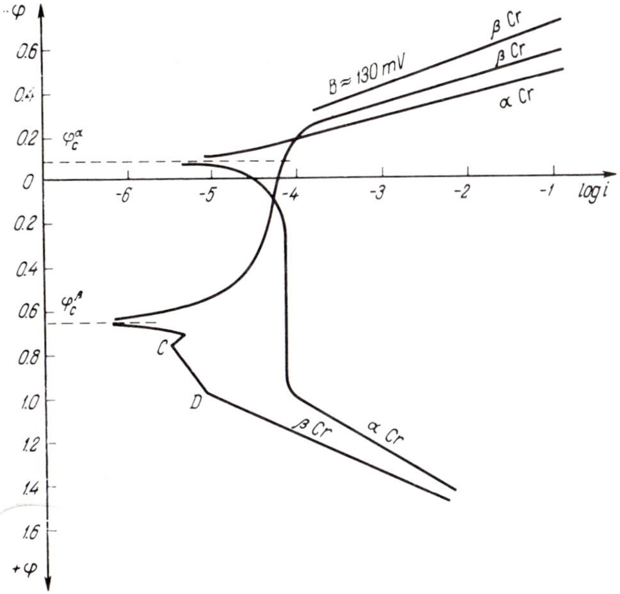

Fig. 3. Differing electrochemical behavior of the two forms of chromium. From Nenov, I.P., Christov, S.G., Raichev, R., and Georgiev, Z., *Electrochimica Acta,* 12, 1537-1544, 1967 (in German). Reproduced with permission.

Nenov, Christov, Raichev and Georgiev [1], and Kristov, Nenov and Raichev [2], report on the differences in the electrochemical behavior of the two forms. Figure 3 shows their results on samples of chromium activated by etching with concentrated hydrochloric acid. The negative potentials are cathodic and the positive anodic. It can be noted that the β chromium (hexagonal) is quite different from the α chromium (cubic). The β chromium passivates more strongly and has a greater corrosion resistance. The β chromium has a hydrogen overvoltage 0.089 volts higher than α chromium under these conditions. In other tests where the samples are activated by cathodic treatment at 20°C and 25 ma/cm^2 (25 asf) in 1N K_2SO_4 for 6 hours, the hydrogen overvoltage is 0.190 volts higher on β chromium than on α chromium.

While cold chromium or β chromium has some desirable properties, it should be mentioned at this point as a further precaution that it is dangerous to attempt to use it for heavy hard chromium deposits for wear resistance, due to its intrinsically poor adhesion. One of the great advantages of hard chromium plating from warm solutions is that the adhesion is almost always of a very high order, unless perchance the current is interrupted and the second layer of chromium tends to flake off of the first.

The high current efficiency with which cold chromium is produced is very attractive but difficult to utilize commercially. It is commonly two or three times as high as the current efficiency in warm solutions, generally in the range of 30-40% compared to about 10-20% in heated baths. Figure 4 shows some data from D.G. Rogers and A.A. Burr extending down to 25°C [3]; it can be noted that the efficiencies at 25°C are all distinctly higher than those at higher temperatures.

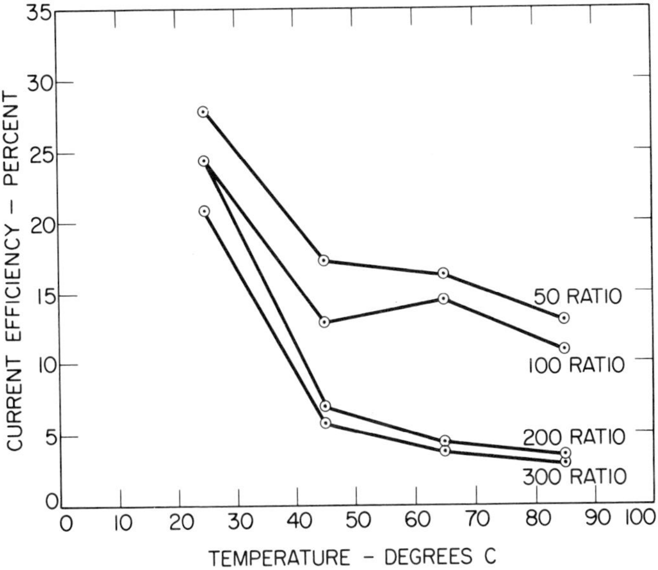

Fig. 4. Average current efficiencies of chromium deposition in cold and warm solutions at current densities over 15.6 amp/dm^2 from data of Rogers and Burr [3].

References

1. Nenov, I.P., Christov, S.G., Raichev, R., and Georgiev, Z., Electrochimica Acta **12**, 1537-1544, 1967 (in German).
2. Khristov, S.G., Nenov, I.P., and Raichev, R.G., Proceedings Third International Congress on Metallic Corrosion, Moscow, 1966, Volume 1, pages 389-395 (in English). Published 1969 by Swets & Zeitlinger, Amsterdam.
3. Rogers, D.G., and Burr, A.A., J. Electrochem. Soc. **97**, 92-98, 1950.

Charles H. Eldridge in 1943.

Kevie W. Schwartz (standing) with Robert H. Dudley in 1944.

Vernon C. Burr in 1944.

Theodore H. Webersinn in 1944.

(Standing left to right) J.E. Stareck, H. Mahlstedt, Olga Behrman Polushkin, H.D. McLeese, and E.L. Kries; (seated) T.G. Coyle, G. Dubpernell, R.O. Loengard and J.M. Hosdowich in 1953.

(Left to right) G. Dubpernell, E.J. Seyb, H. Chessin, M.B. Kalt and F. Passal in 1974.

First pilot plant of Chromium Corporation of America and United Chromium, Incorporated in Waterbury, Connecticut, about 1928.

Large film casting wheel chromium plated at the Eastman Kodak Company in 1926.

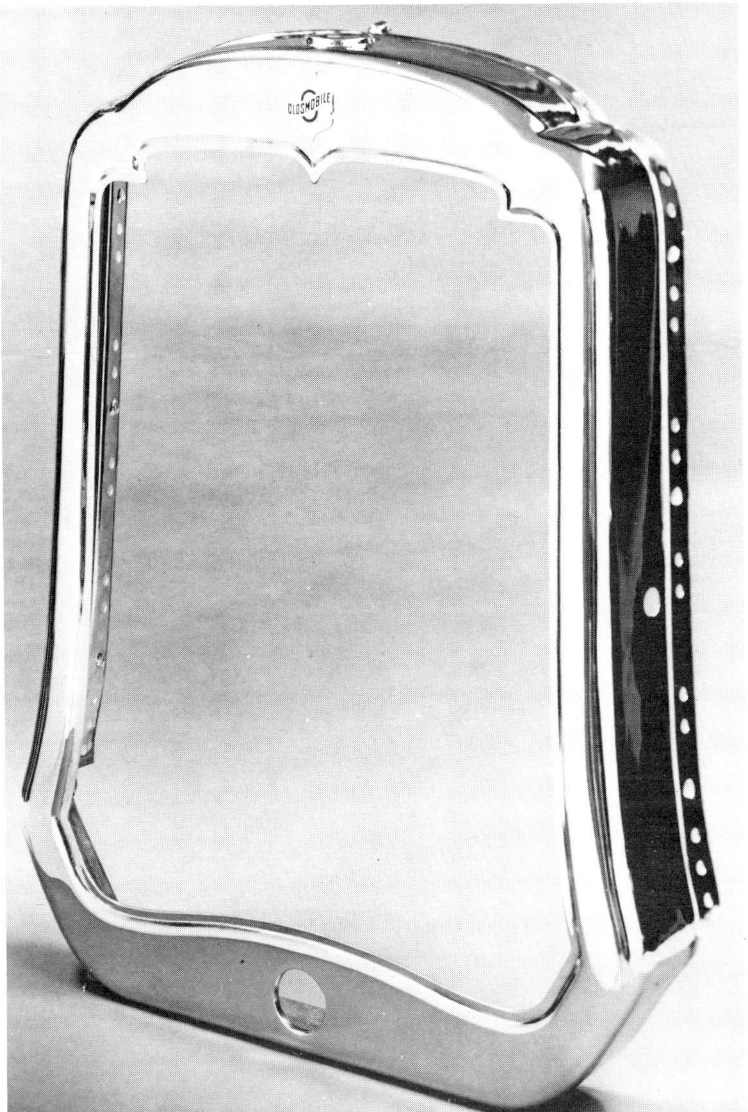

Radiator shell for 1926 Oldsmobile, the first major large-surface use of decorative chromium for an American car.

Corner of Chromium Corporation of America plant in Waterbury, from a 1937 catalog. A large roll is being chromium plated by rotation partially submersed, with two motor-generator sets to provide the required high current. This plant opened in 1926.

Large drying roll with working surface chromium plated, ground and polished at the Chromium Corporation of America. From a 1948 catalog.

Large 6' x 12' roll being chromium plated in the Chromium Corporation of America plant. From an article in Plating in January, 1949.

Men inspecting a heavy cylinder chromium plated on the inside at Chromium Corporation of America. From an article in Plating in January, 1949.

E. Knill points to a roll chromium plated and polished to mirror brightness in the plant of the Chromium Corporation of America. Next to him in this 1964 photo is Ralph Colter, the plant supervisor for many decades.

4
The Catalyst Balance

Pure chromic acid solutions are little different from sulfuric acid solutions when subjected to electrolysis – only hydrogen is evolved at the cathode and there is ordinarily no other visible reaction. If now more than about 0.4% by weight of the chromic acid of sulfate ion or sulfuric acid is added, the chromic acid is reduced at the cathode to trivalent chromium at low current densities, and this trivalent chromium dissolves in the chromic acid solution turning it dark-colored or black. At higher current densities hydrogen begins to be evolved and chromium starts to be deposited, along with some reduction of the solution simultaneously. The reaction is the same if the same amount of sulfate ion is added in the form of any soluble sulfate instead of sulfuric acid. The sulfate ion is unaffected by either the anode or the cathode reactions, and remains in the solution. It is therefore regarded as a catalyst.

The behavior is in part illustrated by the current efficiency curve in Fig. 5 for a solution containing 250 g/l of pure chromic acid, adapted from the work of Haring and Barrows in 1927 [1]. As the sulfate ion or sulfuric acid concentration is increased in small

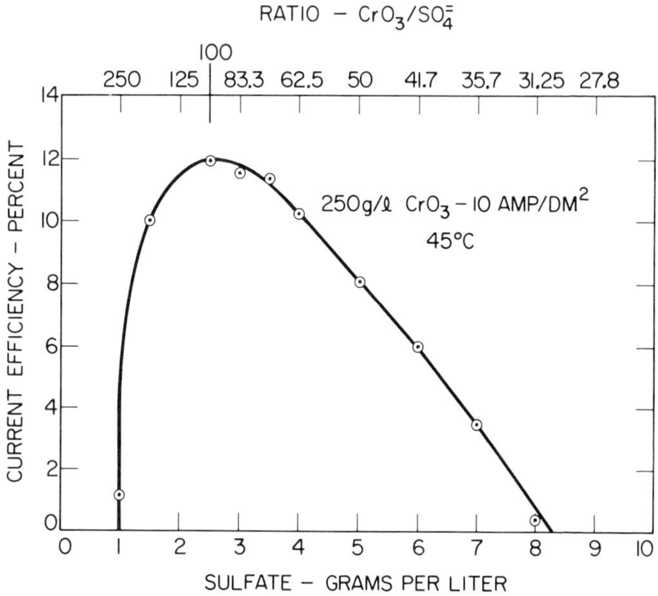

Fig. 5. Haring and Barrows curve of current efficiency versus sulfate concentration showing the maximum at about 100:1 ratio of $CrO_3:SO_4^{--}$.

increments we find that at first the quality of the chromium deposit and the current efficiency increase rapidly, reaching a maximum at about 2.5 g/l sulfate (100/1 ratio). This corresponds to a maximum in the throwing power and covering power also. Further increases in sulfate (decreasing ratio) give a brighter plate but start to decrease the current efficiency slightly. This is the normal working range of the solution up to about 3.125 g/l $SO_4^=$ or 80 ratio.

Beyond this point the efficiency and throwing power decrease more rapidly, and the solution is said to be "oversulfated." Ultimately, when more sulfate is added, no deposit is obtained at all unless the current density is increased. At low current densities the current is all consumed in the formation of trivalent chromium. There is a tendency to pulsating gassing, or alternating periods of chromium deposition and of reduction to trivalent chromium.

The solution containing 250 g/l CrO_3 and 2.5 g/l $SO_4^=$ or H_2SO_4 has come to be adopted very widely all over the world as the standard or average, all-purpose chromium plating bath. It was first disclosed by Fink [2] in 1926 who said: "The amount of stable radicals (catalytic agent) in the bath should approximate 2.5 grams – of sulphate radical – per liter of a solution containing 250 grams per liter of chromic acid." Using Fink's principle of a ratio of CrO_3 to $SO_4^=$ of about 100 to 1, chromium plating baths are in wide use ranging in concentration from about 50 g/l to about 400 g/l.

The standard solution of 250 g/l CrO_3 and 2.5 g/l of $SO_4^=$ is often incorrectly called "Sargent's" solution, a custom which was originated by opponents to the enforcement of Fink's patents in America and of Liebreich's patent in Europe [3]. A more recent account of Sargent's work has been published [4]. Sargent used trivalent chromium salts indiscriminately in his baths to adjust the concentration of trivalent chromium. He did not distinguish between chromic sulfate and chromic chloride, for example.

Many catalysts other than sulfate ion have been proposed from time to time, but the only ones to receive noteworthy commercial use are fluoride and silico-fluoride. Even these give dark gray or "black"

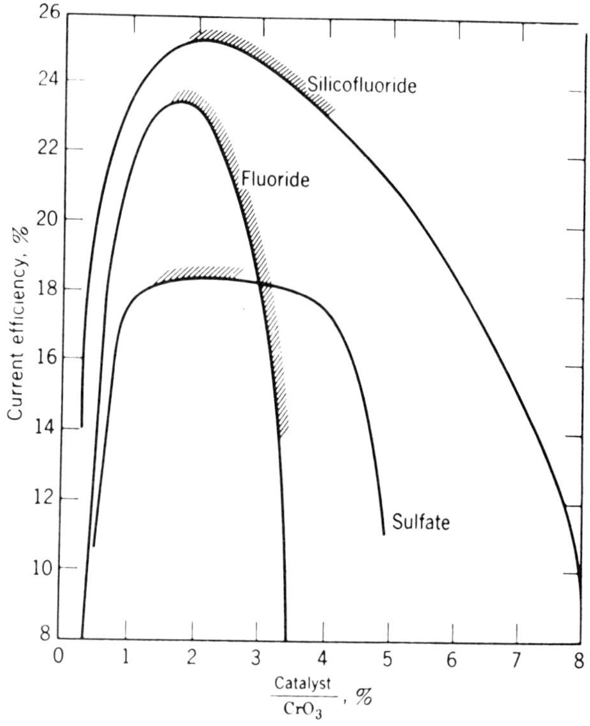

Fig. 6. Bilfinger's curves of the current efficiency of sulfate, fluoride and silicofluoride in 250 g/1 CrO_3 baths at 55°C (131°F) with 50 amp/dm^2 (460 amp/ft^2). From Lowenheim, F.A. (Ed.), *Modern Electroplating*, 3rd ed., John Wiley, 1974. Reproduced with permission.

chromium deposits when used alone with pure chromic acid [5-9], and the presence of a minimal amount of sulfate, about 0.3 g/l or more, is required to get normal chromium deposition.

An approximate indication of the relative behavior of these three catalysts has been given by Bilfinger

[10] as shown in Fig. 6. The contrary indication of many workers of fluoride as a weaker catalyst than sulfate (see Haring and Barrows [1] for example) is due to making the tests in glass containers. The fluoride forms hydrofluoric acid in the strong chromic acid medium, and this rapidly reacts with glass or any other source of silica to give silicofluoride, which is only about one-half as strong a catalyst as sulfate. The simple fluoride ion is a much stronger catalyst than sulfate, about four times as strong [11].

Chloride is also a relatively common catalyst, either by accident or intent. Fink called it an unstable catalyst because it is readily liberated as chlorine gas at the anodes of lead dioxide. This behavior is used to purify solutions from the unwanted catalyst. The final smaller amounts or traces of chloride can be precipitated as insoluble silver chloride by addition of the stoichiometric quantity of silver oxide, carbonate or sulfate.

Chloride is about eight times as strong a catalyst as sulfate, and rather small amounts will decrease the throwing power in decorative plating. It is thus generally regarded as an impurity rather than as another catalyst. It also tends to give a slightly dull plate, less bright than that with sulfate or fluoride, and this likewise classifies it as an impurity. A further disadvantage is a strong tendency to cause etching of the basis metal wherever the substrate is not protected by chromium plate or other means. A maximum upper limit of chloride ion that may be more or less successfully tolerated or allowed for, is about 0.1 g/l, and generally somewhat less. Nevertheless, small amounts can be taken into consideration in adjusting the catalyst balance of the solution, by using corre-

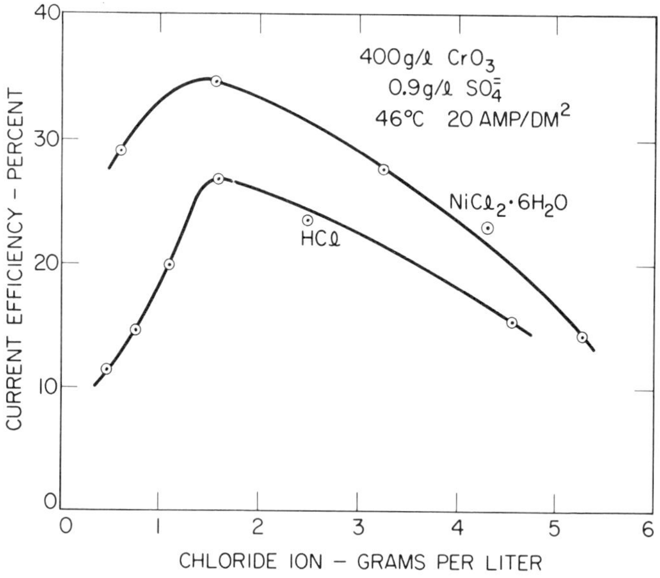

Fig. 7. Mankowich's curves of current efficiency of solutions with chloride catalyst.

spondingly less of sulfate or fluoride until the chloride is eliminated, and a new balance arrived at.

While the use of chloride as a catalyst is rare, it is not unknown. Thus Sargent [12] used chromic chloride additions instead of chromic sulfate additions in one series of tests, and found the chloride to give only slightly less good results than the sulfate for current efficiency and quality of deposit.

In 1927 Mankowich investigated chloride additions to 400 g/l CrO_3 solutions containing varying amounts of sulfate [13]. He obtained the best results with the least amount of sulfate tried, 0.9 g/l. His results for current efficiency are given in Fig. 7, using

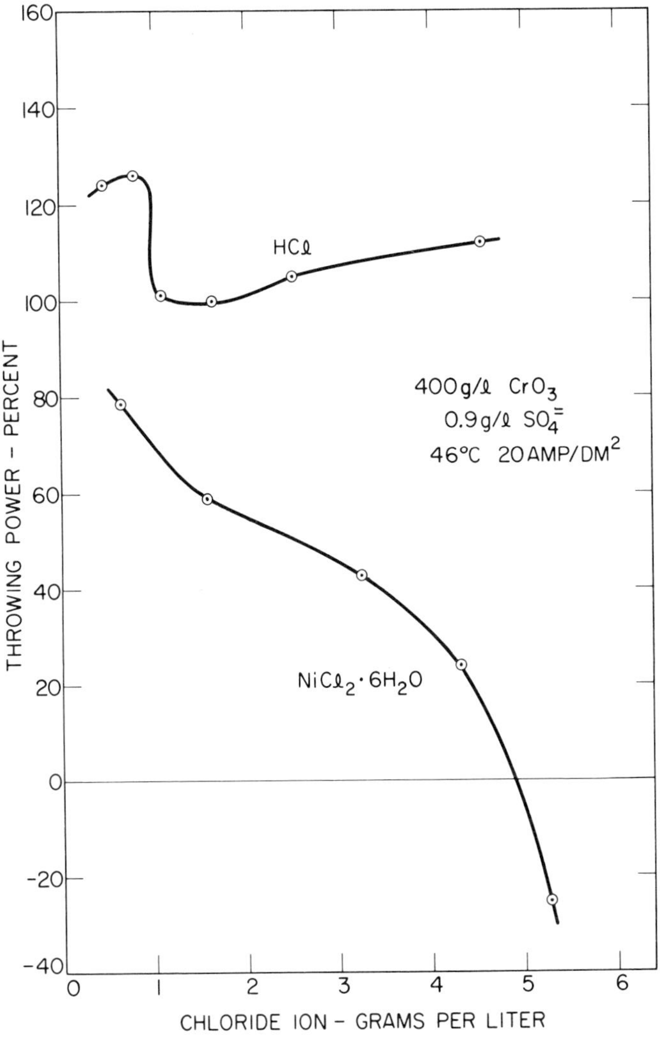

Fig. 8. Mankowich's curves of throwing power of solutions with chloride catalyst.

the average chloride content of the solution in 30 minute runs, based upon analysis of the solution at the beginning and end of each run. The conditions were 46°C and 20 amps/sq dm. While $NiCl_2 \cdot 6H_2O$ additions gave higher current efficiencies than HCl additions, the HCl additions gave a radically higher throwing power of over 100%, that is, deposits of chromium thicker on the low current density cathode further from the anodes, than on the high current density cathode nearer to the anodes, as shown in Fig. 8. This behavior would indicate that the chloride is tightly combined in a complex ion in the nickel chloride.

These results are only of theoretical interest since the deposits were all of a dull grey or smoky appearance, and poisonous chlorine gas is given off at the anode in large quantities. The trivalent chromium content of the solutions was not determined.

Taubald [14-15] gives current efficiencies at 40 amp/sq. dm and 30°C for a solution containing 400 g/l chromic acid and varying amounts of chloride added from about 1.1 to 11 g/l. His results are in good agreement with those of Mankowich.

Recently chloride-catalyzed solutions of low concentration have been tested for the production of chromium plated steel strip for making food containers [16-17]. Carter [18-19] employs a 100 g/l CrO_3 solution with a ratio of CrO_3 to chloride ion of 750:1; in other words the chloride ion seems about 7.5 times as strong as sulfate. It is interesting that Hull and Winters [20] also found chloride to be 7.5 times as strong a catalyst as sulfate.

Stareck, Seyb, Johnson, and Rowan [21] found a very strong effect of chloride to promote cracking and cut down the thickness to which bright crack-free chromium plate could be produced. In their limited range of conditions they consider chloride to be 14 times as strong a catalyst as sulfate.

Solov'eva, Lapshina and Vagramyan [22] and Kazakov and Vagramyan [23] have reported good current efficiencies in solutions catalyzed with chloride alone, but did not describe their deposits. They indicate chloride to be a weaker catalyst than sulfate, but since this is not true it must be assumed that most of the chloride added was quickly lost as chlorine gas at the anode, and thus failed to exert its catalytic effect at the cathode. It should also be noted that lead dioxide coatings on anodes from previous electrolysis are capable of evolving chlorine gas from strongly acid solutions chemically, without the passage of current. Thus, chloride must be regarded as an unstable catalyst, and its concentration should be determined before and after each experiment, unless other than lead dioxide anodes are used which possibly do not evolve chlorine gas so rapidly.

Griffin [24] also found that chloride was a catalyst for chromium deposition somewhat like sulfate.

An example of the positive throwing power noted by Mankowich with chloride additions was encountered in plant practice in 1949, in the heavy hard chromium plating of "pig tail" thread guides. These had a normal bright chromium plate, accompanied by a welt of dull chromium plate on the low current density areas which appeared to be thicker than the bright chromium on the high current density areas. The cross section of one of these thread guides at about 25 magnifications is shown in Fig. 9. It appears

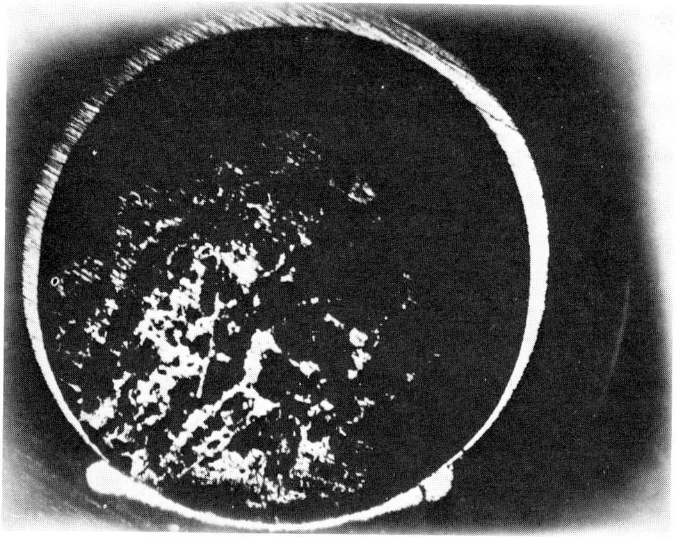

Fig. 9. Cross section of steel thread guide 1/8″ diameter at 25X showing normal hard chromium plate and welt of thick, dull chromium on low current density area due to chloride in solution.

likely that the bright chromium is deposited under a semiconducting film of chromic oxide as has been postulated by some authors, and that this resistance shifts part of the current to the low current density areas. Chemical analysis of the solution in this case showed it to contain approximately:

Chromic acid . . . . 240 g/l
Sulfate . . . . 3.2 g/l
Chloride . . . . 0.7 g/l

In summary, it might be said that the approximate relative strengths of the common catalysts for chromium baths is in the ratio of 8:4:1:0.5 for chloride, fluoride, sulfate, and silicofluoride respectively. While the catalysts are to a degree interchangeable, the figures should be taken as a crude guide only, taking into consideration that the specific chemical properties of each catalyst will dominate if it is present in large proportion.

It has sometimes been thought that the effectiveness of the different catalyst anions might be related to their chemical equivalent weights, but no such relationship appears discernable. Thus, taking one half of the sulfate ion as one, the ratios of Cl^-, F^-, 1/2 SO_4^{--}, and 1/2 SiF_6^{--} are as 35.46:19:48:71 or 0.74:0.396:1:1.48, and these ratios do not seem to have any relationship to the above ratios of 8:4:1:0.5.

Some examples of baths containing catalysts having a total effect similar to the standard bath at 100:1 ratio (No. 1) may be helpful:

Table 2

1. CrO_3 – 250 g/l
   $SO_4^=$ – 2.5 g/l
   $$\text{ratio} = \frac{250}{2.5} = \frac{100}{1}$$

2. CrO_3 – 250 g/l
   Cl^- – 0.3125 g/l
   $$\text{ratio} = \frac{250}{0.3125 \times 8} = \frac{100}{1}$$

3. CrO_3 – 250 g/l
   F^- – 0.625 g/l
   $$\text{ratio} = \frac{250}{0.625 \times 4} = \frac{100}{1}$$

4. CrO_3 – 250 g/l
   SiF_6 – 5.0 g/l
   $$\text{ratio} = \frac{250}{5 \times 0.5} = \frac{100}{1}$$

5. CrO_3 – 250 g/l
   $SO_4^=$ – 1.0 g/l
   SiF_6 – 3.0 g/l
   $$\text{ratio} = \frac{250}{1 + (3\times0.5)} = \frac{100}{1}$$

6. CrO_3 – 250 g/l
   SO_4 – 1.7 g/l
   Cl^- – 0.1 g/l
   $$\text{ratio} = \frac{250}{1.7 + (0.1\times8)} = \frac{100}{1}$$

Baths No. 2, 3 and 4 above are extremes and should not be considered practical working baths. Bath No. 5 is somewhat similar to some self-regulating baths in common use. Bath No. 6 is an ordinary chromium bath contaminated with 0.1 g/l chloride. If there was no further contamination, the amount of sulfate would have to be increased as the chlorine was electrolyzed out at the anode.

The above baths are given on a theoretical basis, as if made up with chemically pure chromic acid. If made up with commercial chromic acid containing a variable small percentage of sulfate of the order of 0.1 to 0.2%, this has to be determined by chemical analysis and included in the calculations.

While the ratio of $CrO_3:SO_4^{=}$ of 100:1 has been a convenient and helpful standard, it is not always the optimum, and the use of this ratio has been rather highly overworked. This amount of sulfate is rather minimal under most circumstances, and tends to decrease due to drag out, so that "undersulfated" solutions requiring the addition of more sulfate have been a common occurrence. It would be better to use a ratio of 80 or 90 as the standard for most purposes.

About the only exception is in barrel chromium plating where the current densities and temperatures are quite low in the complex fluoride catalyzed solutions used, and a ratio of about 110-115 appears optimum.

References

1. Haring, H.E., and Barrows, W.P. "Electrodeposition of Chromium From Chromic Acid Baths," Technologic Papers of the Bureau of Standards Vol. **21**, No. 346, 413-449, June 10, 1927, Washington D.C.: U.S. Government Printing Office.
2. Fink, C.G., assignor to Chemical Treatment Company, U.S. 1,581,188 – April 20, 1926.
3. Dubpernell, G., Plating **47**, 35-53, 1960.
4. Dubpernell, G., Plating **59**, 638, 640, 643, 1972.
5. Drobantseva, N.T., Vestnik Mashinostroeniya **36** (5), 68, 1955; Brutcher Translation No. 3798; see C.A. **50**: 12968.
6. Drobantseva, N.T., and Sysoev, A.N., Zhur. Prikladnoi Khim. **29** (4), 589-595, 1956; J. Applied Chem. U.S.S.R. **29**, 647-652, 1956 (in English); see C.A. **50**: 15286.
7. Drobantseva, N.T., and Sysoev, A.N., Russian pat. 101,526 – Dec. 31, 1955; see C.A. **51**: 13625.
8. Graham, A.K., Proc. Am. Electroplaters' Soc. **46**, 61-63, 362, 1959.
9. Pinkerton, H.L., to Allied Chem. Corp., U.S. 2,985,567 – May 23, 1961; Can. 624,887 – Aug. 1, 1961.
10. Bilfinger, R. "Das Harverchromungs – Verfahren" (The Hard Chromium Plating Process), page

84, Leipzig: 1939; 2nd edition, pp. 39, 134-135, Ann Arbor, Michigan: Edwards Brothers, 1942 . See also Hood, T.A., Plating Notes (Australia), **4**, 31, April 1952; Metal Finishing **50**, 103, June, 1952.

11. Seyb, E.J., Jr., Woehrle, R.E., and Neitzel, J.G., to M&T Chemicals Inc., U.S. 3,498,892 – March 3, 1970.
12. Sargent, G.J., Trans. Am. Electrochem. Soc. **37**, 479-497, 1920.
13. Mankowich, A. "The Effect of the Chloride Anion on the Throwing Power and Current Efficiency of Chromic Acid Plating Solutions," unpublished report No. 12 of the Chromium Corporation of America, 21 pages, December 8, 1927.
14. Taubald, K., Galvanotechnik **64** (3), 184-190, March 1973.
15. Taubald, K. "Mechanism of the Electrodeposition of Chromium-Selenium Alloys" (in German), Dissertation Dresden, 1970.
16. Smith, J.R., and Bray, J.A., to National Steel Corporation, U.S. 3,475,295 – October 28, 1969.
17. Serra, G., to United States Steel Corporation, U.S. 3,532,608 – October 6, 1970.
18. Carter, W.A., Jr., and Lee, H.H., to Inland Steel Company, U.S. 3,567,599 – March 2, 1971.
19. Carter, W.A., Jr., to Inland Steel Company, U.S. 3,616,303 – October 26, 1971.
20. Hull, R.O., and Winters, J.B., Proc. Am. Electroplaters' Soc. **36**, 93-101, 1949.
21. Stareck, J.E., Seyb, E.J., Jr., Johnson, A.A., and Rowan, W.H., to Metal and Thermit Corporation, U.S. 2,952,590 – September 13, 1960.

22. Solov'eva, Z.A., Lapshina, A.E., and Vagramyan, A.T., pages 358-365 in the book edited by Fedot'ev, N.P. "Protective Metallic Coatings," English translation prepared by the Israel Program for Scientific Translations, 1968; available from Clearinghouse for Federal Scientific and Technical Information, Springfield, Va. 22151, under No. TT67-51361.
23. Kazakov, V.A., and Vagramyan, A.T., Elektrokhimiya 2 (2), 258-261, 1966; Soviet Electrochemistry 2 (2), 241-243, (1966).
24. Griffin, J.L., Plating **53**, 196-203, (1966).

5
Empirical Tests for Catalyst Concentration and Bath Balance

The electrodepositon of chromium from chromic acid solutions is so sensitive to, and controlled by, such small concentrations of catalyst anions that it was recognized at an early date that the characteristics of the process itself could be used for its control, and for the determination of catalyst concentration. Thus, Fink and Eldridge found in 1924 that the approximate purity of a shipment of chromic acid could be checked quickly and conveniently by means of a "no plate-plate test" on a small amount of solution without any additions. If the test solution remained clear and gave steady gassing and only a brown film on the brass strip cathode at a high current density, it was known that the chromic acid contained less than 0.4% sulfate and was of a purity suitable for chromium plating. On the other hand, if the solution turned dark due to reduction, a higher current density was required to get gassing, the gassing was intermittent, the brass strip etched at the solution line, and some chromium plate was obtained at high current density points, it was known that the chromic acid was too impure and contained too much sulfate to be used for chromium plating.

Buffed copper or brass strips were plated under standard conditions partly immersed in the solution, to determine the current efficiency and appearance of the plate, for control purposes. If these were normal, nothing needed to be done. If the deposit was rough and grayish with patches of brown oxide coating, and had a low current efficiency, it was evidence of "under-sulfation," and more sulfate needed to be added to get a clear bright plate. If the strip was bright but poorly covered at the solution line with a "meniscus" and etching of the brass or copper, and required a higher current density to get started gassing and plating, and to avoid periodic gassing, then it was known that the solution was "over-sulfated," and would require the addition of some barium carbonate or pure chromic acid to bring it back into balance.

Some operators could control the composition of their solutions fairly well in this way, but the best practice was found to be the analysis of the solution for both hexavalent chromium and sulfate, even though the latter analysis was difficult in the early years and required setting overnight in a warm place to get all of the sulfate to precipitate after reducing a sample to the trivalent stage. Willard and Schneidewind found in 1929 [1] that the time for the precipitation of sulfate completely from complex chromium sulfate solution could be reduced to about an hour by precipitating in the presence of acetic acid, which formed a tighter complex with the trivalent chromium and liberated the sulfate.

The variation in the throwing power or covering power of the chromic acid bath was found to be a means for the approximate determination of the sulfate content, or the ratio. Thus, Pinner and Baker [2] used a copper strip cathode with a right angle bend in order to estimate the ratio of chromic acid to sulfate. They used an optimum ratio of chromic acid molarity to sulfate normality of 50 to 1 instead of the more commonly used weight ratio of 100 to 1, but there is no difference. They felt that their test would take into consideration other acid radicals than sulfate which might be present, but did not make any quantitative comparison of its accuracy as compared to chemical analysis.

Pinner and Baker presented a second paper [3] with the same data, and emphasized that their plating test was being used successfully as a means of plant control, to take the place of chemical analysis. However, this type of test has not replaced analysis to any great extent, but rather is used for supplementary information and checking.

While bent cathode tests of one type or another are still in common use to judge the plating characteristics of a solution, they are not generally applied to estimating and controlling chemical composition. The Hull cell developed by R.O. Hull [4, 5], inspired by studies of throwing power and current distribution by Busse [6] and Kasper [7], has become highly standardized and widely used. Its use to estimate the sulfate content or bath balance of chromium plating baths was developed by MacIntyre and Hull [8], and Hull and Winters [9, 10]. A German book by Nohse on the use of the Hull cell appeared in two editions [11, 12] with a still more complete English edition in 1966 [13].

A single test on a Hull cell panel plated at an angle to the anode shows the plating characteristics of the solution over a very wide range of current densities. By comparing this panel with standard panels plated under known conditions it is possible to deduce considerable information regarding the composition of the plating bath and its needs for control and maintenance, especially if additional panels are plated to test the effect of known additions to the solution made in the Hull cell itself.

In the case of decorative chromium plating solutions, the sulfate content can probably be estimated to within about ± 10% or ± 15% in this way. What is more important, the test gives an indication of the actual performance of the bath, and the total of all catalysts present. Thus, as described by Corfe in the Nohse book [13], as little as 0.01 g./l. of chloride in a chromium bath can be detected by its effect on the coverage of the panel and the etching on low current density areas, if a buffed brass panel is used. Various types of current distribution and plating tests are comprehensively described by Luce [14].

It is felt [10, 13] that a Hull cell test can be used to check the performance of a mixed catalyst or self-regulating chromium bath. While this is true, not much can be learned regarding the individual catalyst concentrations without chemical analyses.

Another common go-between intermediate between a carefully conducted analysis for sulfate and a quick plant test such as the Hull cell test is the centrifugal determination of sulfate introduced by the State Manufacturing Company and the Kocour Company in 1929 [15]. The use of a centrifuge permits a sulfate determination by precipitation of barium sulfate in the presence of chromic acid in about ten minutes. This method is widely employed, and is quite useful in the hands of a careful operator. It is an empirical method depending upon compensating errors and careful standardization of each step. The results should be checked on a known solution occasionally, as misleading results have sometimes been obtained. Foulke [14] gives a good description of the method.

Bilfinger [16, 17] gives procedures both in the presence and absence of silicofluoride. If silicofluoride is present it is destroyed by heating with ammonia, and the silica which precipitates is filtered off before precipitating the sulfate as usual on an aliquot part of the filtrate. This would be much slower than dealing with sulfate alone. Some Japanese workers [18] also report high results in the presence of silicofluoride due to coprecipitation of barium silicofluoride.

Blum and Hogaboom [19] give the accuracy of the centrifuge method as ± 10% which is barely satisfactory or useful for control purposes. However, good standardization of each step of the procedure probably gives considerably better results than this, more nearly in the range of ± 5% mentioned by Langford and Parker [20]. One possibly critical point in the procedure is the use of an inexpensive, hand-cranked centrifuge, which could lead to variations. A motor-driven centrifuge could probably contribute greater reproducibility.

Other empirical tests for sulfate in chromium plating baths were developed in the early years, 1925-1929, but were not published or placed in common use. It was found that the amount of reduction or of trivalent chromium formed at the cathode was proportional to the sulfate concentra-

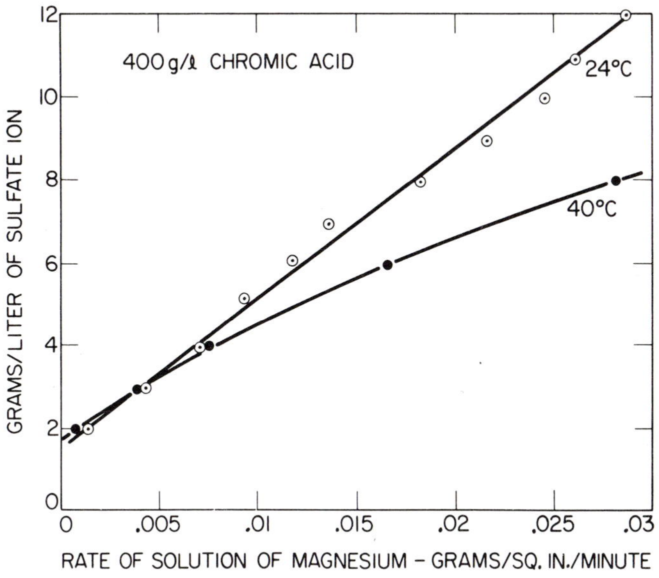

Fig. 10. Dubpernell's determinations of the rate of solution of magnesium in 400 g/l CrO_3 at 24°C and 40°C with varying sulfate concentration.

tion, but this was difficult to determine. More promisingly, magnesium metal was found to dissolve in the bath at a rate which varied almost linearly with the sulfate content, beginning at a sulfate concentration corresponding to about a ratio of 200:1 of chromic acid to sulfate, as shown in Fig. 10. The magnesium is substantially insoluble in pure chromic acid or in chromic acid containing small amounts of sulfate.

From this information a magnesium ribbon test was developed which had potentialities and has been briefly described by Soderberg [21]. Figure 11 shows one type of tester which was used. A short piece of

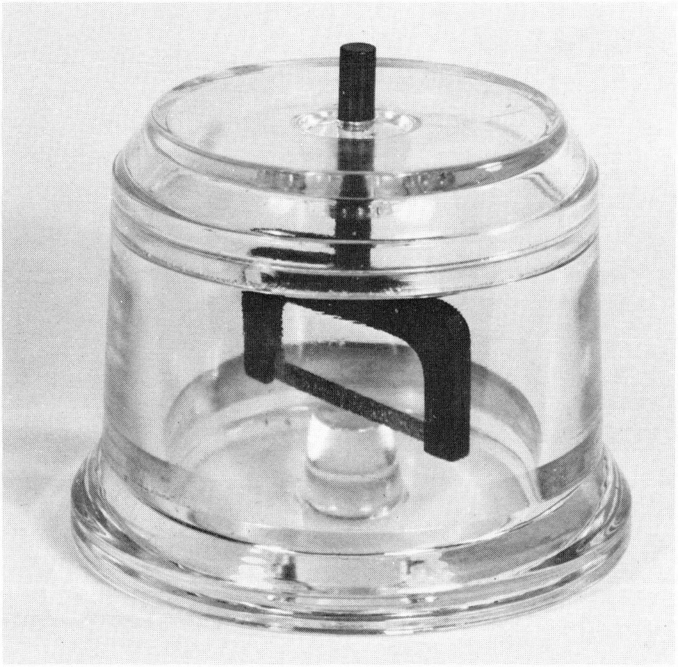

Fig. 11. Magnesium ribbon tester for the determination of the sulfate content or ratio of chromium plating baths.

standard magnesium ribbon 0.125″×0.006″ with 1.5″ length exposed was placed in a plastic frame holder and mounted on a glass pedestal as shown. About 150 cc. of plating solution to be tested is cooled to a room temperature of about 25°C and placed in the container with the magnesium ribbon. The time is noted when the holder drops due to the dissolving through of the magnesium ribbon, and the sulfate concentration or ratio of chromic acid to sulfate is obtained from a calibration chart such as Fig. 12. Generally a time of about eleven minutes is required for 100 ratio solutions in the range of 250 to

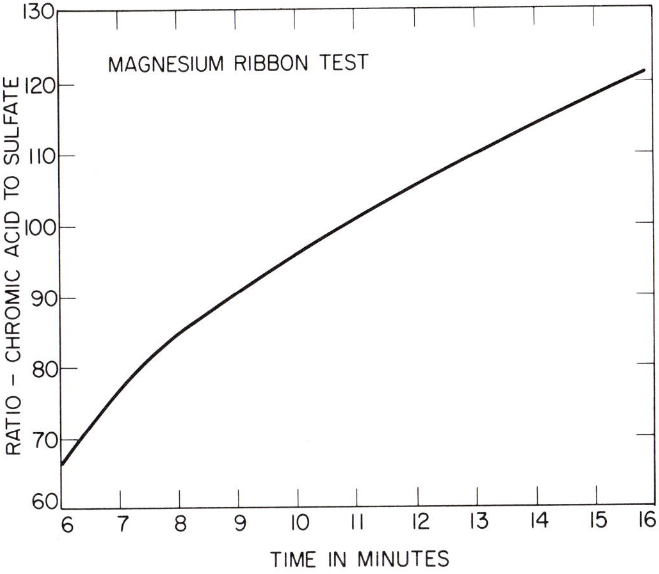

Fig. 12. Calibration chart for magnesium ribbon test for $CrO_3:SO_4^{=}$ ratio.

400 g./l. CrO_3. The test is remarkably insensitive to moderate variations of temperature and chromic acid concentration. However, in practice it only seemed accurate to about ± 10% or more, and was dropped in favor of chemical analyses.

An interesting feature of this test is that in freshly made up solutions it can distinguish between sulfate added as sulfuric acid or sodium sulfate, and sulfate added in a complex salt such as chromium sulfate, whereas no difference is found in plating tests. When the sulfate is added as chromium sulfate the magnesium ribbon takes longer than eleven minutes to dissolve, presumably because some of the sulfate is

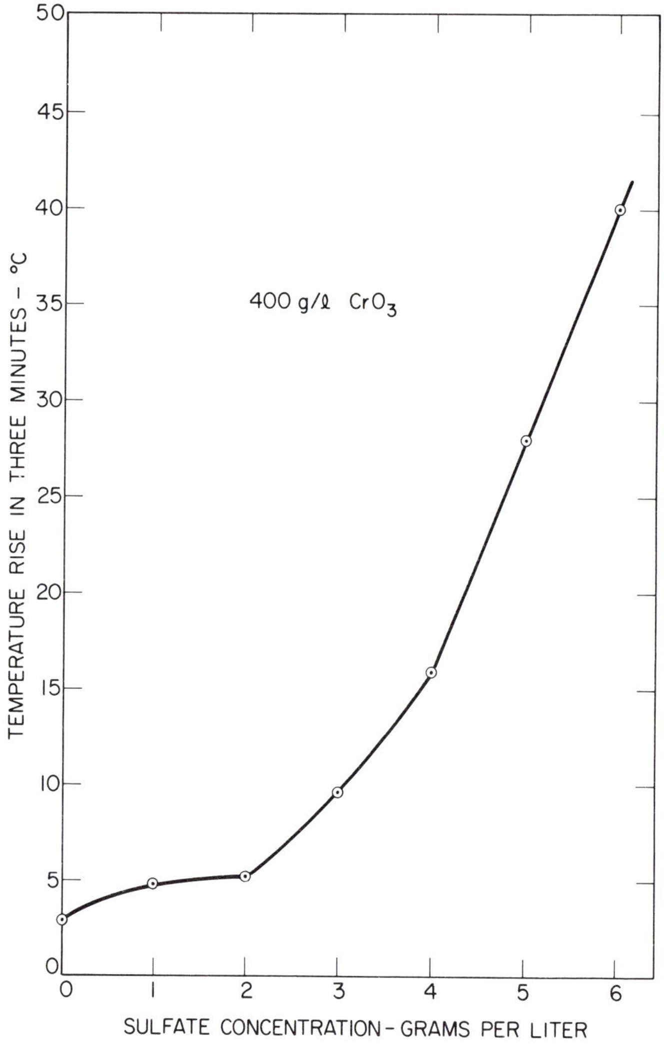

Fig. 13. Calibration chart for zinc dust test for sulfate concentration of 400 g/l CrO_3 solutions.

tightly combined in a complex ion. However, if the test is repeated on the same small volume of solution, or the solution is first electrolyzed for a few minutes, the standard time of eleven minutes is obtained, when using a 100 ratio composition.

While zinc dissolves more slowly in chromium plating solution than magnesium, a faster test for sulfate was developed by using the standard zinc dust produced by condensing zinc vapor in a non-oxidizing atmosphere. The rise in temperature in three minutes when seven grams or 5 c.c. of zinc dust is agitated with 25 c.c. of chromium plating solution in a 50 c.c. flask at 25°C is an index of the sulfate content. An experimental curve for 400 g./l. CrO_3 solution is shown in Fig. 13. The starting temperature is critical and should be 25°C±1°C or better. Care should be used in testing solutions with high sulfate content, as the reaction becomes violent and the sample boils over.

This test is perhaps faster and more convenient than the magnesium ribbon test, but it was never investigated in great detail. A later investigation by A.A. Johnson in 1957 showed a temperature rise in three minutes of 14°C for 25 c.c. of 200 g./l. CrO_3 solution containing 2 g./l. $SO_4^=$, compared to the 16°C temperature rise found above for 400 g./l. CrO_3 solution containing 4 g./l. $SO_4^=$, a good confirmation of the general nature of the test. Johnson also tried the effect of various amounts of silicofluoride catalyst, and found rather startlingly that there was essentially no reaction between 200 g./l. CrO_3 solution and zinc dust with silicofluoride additions, and that the test was apparently capable of determining sulfate in the presence of silicofluroride.

Fig. 14. Pfanhauser's first apparatus for the "potentiometric" determination of sulfate or $CrO_3:SO_4^{--}$ ratio. From Pfanhauser, W., *Verchromungs-Technik,* Langbein-Pfanhauser Werke, 1931. Reproduced with permission.

A brief investigation was made of the use of organic reducing agents instead of zinc dust. Citric acid showed some promise, but still was rather erratic and uncontrollable in its reaction. Probably oxalic acid or some other organic would react smoothly and be more reproducible than zinc dust.

Perhaps the most important empirical test for the sulfate content of chromium plating baths is the "potentiometric" method published by Pfanhauser in 1931 [22] and covered by a German design patent in 1933 [23]. This is based upon a noteworthy investigation of the theory and practice of chromium plating by Pfanhauser and Elssner [24], who noted that the current density at which chromium started to deposit varied with the sulfate concentration.

The apparatus required is shown in Fig. 14 [22],

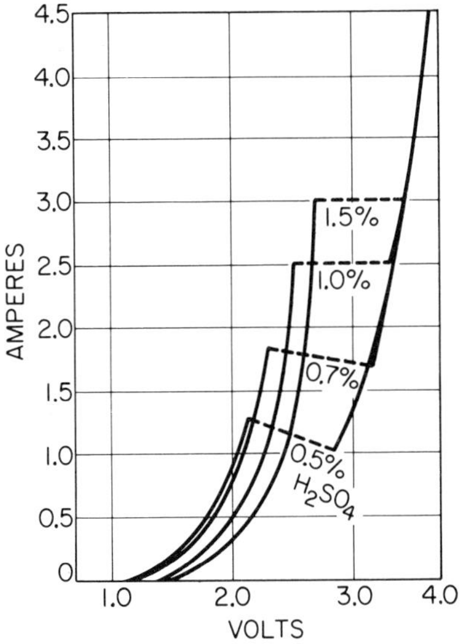

Fig. 15. Bilfinger's curves of results with the L.P.W. Sulfometer. 1.0% H_2SO_4 corresponds to 100:1 ratio of $CrO_3:SO_4^{--}$. From Bilfinger, R., ***Das Hartverchromungsverfahren,*** **Langbein-Pfanhauser Werke, 1939. Reproduced with permission.**

and is simply a small plating cell with standardized dimensions in which a bright brass cathode 70 X 50 mm is plated between two lead anodes, also 70 X 50 mm. The current density at which chromium deposition starts is determined by slowly increasing the current in small increments of 0.2 amperes until there is a sudden increase in voltage of almost a volt, and the current decreases slightly. This current density is recorded and the amount of sulfate or sulfuric acid as a percentage of the chromic acid concentration can be read off from a calibration curve.

The type of results obtained is shown by Fig. 15

taken from Bilfinger [25], and Fig. 16 gives the calibration curve [26]. It is noteworthy that the percent of sulfuric acid in reference to the chromic acid content of the bath varies linearly with the current density at which chromium begins to deposit. The chromic acid concentration has no effect on the calibration curve, but 260 g./l. CrO_3 solutions used for hard chromium plating are said to not give exact results above about 1.25%, and 400 g./l. CrO_3 solutions for decorative plating are inexact above about 1.4%. The measurements are made at a room tem-

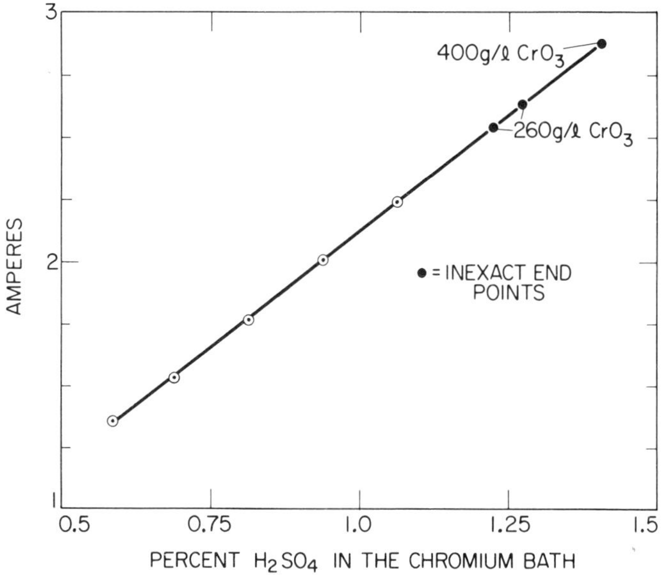

Fig. 16. Calibration curve for Sulfometer testing. Points marked ● are inexact end points. From *Taschenbuch fuer Galvanotechnik* 10th ed., Langbein-Pfanhauser Werke, 1961. Reproduced with permission.

perature of 20 or 21°C±1°C, and are considered accurate to about ±0.05% of sulfuric acid in relation to chromic acid (±5%). A fresh brass cathode has to be used for each test, and the lead anodes are best cleaned or electrolyzed to a reproducible condition for each test also.

This test apparatus came to be known as the L.P.W. Sulfometer, and the ammeter scale was sometimes calibrated directly in percent of sulfuric acid with reference to chromic acid content. The Sulfometer is no longer sold separately, but is only offered as a subsidiary part of a more complete laboratory test set. Holzapfel [27] states that the Sulfometer is not suited to the determination of hydrofluroic acid or fluosilicic acid in chromium baths, without further comment. Bilfinger showed divergent results for hydrofluoric acid and fluosilicic acid in the second and third editions of his book [28, 29].

The results with the L.P.W. Sulfometer method were closely confirmed by Il'in in 1948 [30]. Il'in found that the critical current density upon which the method depends is quite sensitive to temperature. He used 18°C for his tests, and standardized on the resultant somewhat lower critical current densities.

Lovrecek and Bosnjak [31] confirm the accuracy of the L.P.W. Sulfometer method as about ±5% or better, and report that the trivalent chromium content of the bath has practically no influence on the results. They developed a more sophisticated method and apparatus for conducting the tests, using a platinum anode, a brass cathode only 1.54 mm^2 in area, and a high voltage source of current of 67.5 volts. A condenser is connected in parallel with the plating cell so that the charging current can be preset and applied automatically. However, typical charging times are not given and a more detailed investigation would appear desirable. Temperature is claimed to have little influence with this instrumentation.

Stalzer [32] states that the L.P.W. Sulfometer is no longer available, and describes a completely automatic electronic device which he has designed to take its place. His apparatus includes a motor driven potentiometer the scale of which is calibrated to read directly the percent of sulfuric acid in relation to the chromic acid concentration (220-280 g./l.). The method is not sensitive to temperature variations of ±5°C, and corrections can be made for wider variations in temperature or chromic acid concentration. Contaminated baths cause considerable variation and have to be investigated otherwise.

A platinized titanium anode is used by Stalzer with a copper rod cathode of a surface area reproducible to about 1%. After each measurement the copper cathode is discharged through a 10 ohm resistance so that it is immediately ready for re-use. Each measurement requires about 20 seconds. Generally the first measurement is about 20% off due to nonstandard condition of the copper cathode surface, but after this the results are reproducible. Usually three repeat measurements are made, requiring a total time of about one minute, and if the second and third measurements agree they are accepted as the true value. The apparatus is claimed to be as accurate as the centrifuge, but errors are only stated to be within ±10%.

Plumpe [33] describes the adaptation of modern potentiostatic-type equipment to the determination of sulfate in chromium plating solutions, similarly to

Stalzer [32]. Plumpe uses a platinum anode and gold cathode, and an ammeter calibrated directly in percent sulfate in the solution (percent of the chromic acid present). The ammeter shows the peak current registered with a given applied potential, or rate of increase of potential. This peak current is directly proportional to the sulfate percentage, regardless of whether or not it leads to chromium deposition on the gold cathode. The use of the instrument for the automatic control of both sulfate and chromic acid concentration is described, as well as corrections for temperature and concentration. The possible accuracy is not stated.

References

1. Willard, H.H., and Schneidewind, R. Trans. Am. Electrochem. Soc. 56, 33-349, 1929.
2. Pinner, W.L., and Baker, E.M. Trans. Am. Electrochem. Soc. 55, 315-331, 1929; Metal Industry (London) 34, 585-586, 611-612, 1929.
3. Pinner, W.L., and Baker, E.M., Monthly Rev. Am. Electroplaters' Society 16, 16-24, 1929; Brass World 25, 269-272, 1929.
4. Hull, R.O., to E.I. duPont deNemours & Company, U.S. 2, 149, 344 – March 7, 1939.
5. Hull, R.O., Proc. Am. Electroplaters' Soc. 27, 52-60, 1939; Mon. Rev. A.E.S. 26, 753-767, 1939.
6. Busse, O., Zeit. Elektrochemie 38, 783-793, 1932.
7. Kasper, C., Proc. Am. Electroplaters' Soc. 26, 21-32, 1938.

8. MacIntyre, F., and Hull, R.O., Proc. Am. Electroplaters' Soc. 31, 95-105, 1943.
9. Hull, R.O. and Winters, J.B., Proc. Am. Electroplaters' Soc. 36, 93-101, 1949.
10. Hull, R.O. and Winters, J.B., Proc. Am. Electroplaters' Soc. 38, 133-154, 1951.
11. Nohse, W. "The Hull Cell," 60 pages, 1956, Eugen G. Leuze Verlag, Saulgau/Wttbg., Germany.
12. Nohse, W. "The Hull Cell," second edition, 88 pages, 1965, Eugen G. Leuze Verlag, Saulgau/Wurtt., Germany.
13. Nohse, W "The Hull Cell," English edition, 125 pages, 1966, Robert Draper Ltd., Teddington, England.
14. Luce, B.M., in Foulke, D.G., and Crane, F.D. "Electroplaters' Process Control Handbook," 445 pages, 1963, Reinhold Publishing Corporation, New York.
15. State Manufacturing Company, Metal Industry (N.Y.) 27, 394, 491, 1929; Brass World 25, 245, 1929.
16. Bilfinger, R. "The Hard Chromium Plating Process" (in German), second edition, 235 pages, 1942, Herm. Beyer Verlag, Leipzig, Germany.
17. Bilfinger, R. "Handbook of Electrolytic Chromium Plating" (in German), third edition, 331 pages, 1952, Fachbuchverlag Siegfried Schutz, Hannover, Germany.
18. Aotani, K., Kaneko, I., Jin, A., and Kinukawa, S., J. Metal Finishing Soc. Japan 23 (10), 579-584, 1972.
19. Blum, W., and Hogaboom, G.B. "Principles of Eletroplating and Electroforming," third edition,

455 pages, 1949, McGraw Hill Book Company, New York.

20. Langford, K.E., and Parker, J.E. "Analysis of Electroplating and Related Solutions," 4th edition, 498 pages, 1971, Robert Draper Ltd., Teddington, England.
21. Soderberg, K.G., Norsk Galvano – Teknisk Tidsskrift 7, 25-30, 1964; Plating 53, 1262-1265, 1966.
22. Pfanhauser, W. "Verchromungs – Technik" (Chromium Plating Technique), 2nd edition, 229 pages, Leipzig, 1931.
23. Langbein – Pfanhauser-Werke, A.G., D.R.G.M. 1,259,322 – March 14, 1933.
24. Pfanhauser, W., and Elssner, G., Zeit. Elektrochemie **37**, 874-886, 1931.
25. Bilfinger, R. "Das Hartverchromungs – Verfahren" (The Hard Chromium Plating Process), 148 pages, 1939, Leipzig: Herm. Beyer Verlag.
26. Langbein-Pfanhauser Werke A.G. "LPW–Taschenbuch Für Galvanotechnik" (LPW Pocket Book For Electroplating), 10th edition, 927 pages, 1961, Neuss/Rhein: in house publication.
27. Holzapfel, J. "VEM Handbuch Galvanotechnik" (VEM Electroplating Handbook), 2nd edition, 662 pages, 1967, Berlin: VEB Verlag Technik.
28. Bilfinger, R. "Das Hartverchromungs – Verfahren" (The Hard Chromium Plating Process), 2nd edition, 235 pages, 1942, Leipzig: Herm. Beyer Verlag.
29. Bilfinger, R. "Handbuch der elektrolytischen Verchromung" (Handbook of Electrolytic Chromium Plating), 3rd edition, 331 pages, 1952, Hannover: Fachbuchverlag Siegfried Schuetz.

30. Il'in, V.A., Zavodskaya Laboratoriya **14** (11), 1389-1390, 1948: Electroplating **3** (4), 130-131, 1949; Henry Brutcher Translation No. 2280, Altadena, California.
31. Lovrecek, B., and Bosnjak, K., Metalloberflaeche **11** (1), 19-20, 1957.
32. Stalzer, M., Galvanotechnik **59** (10), 815-817, 1968; Galvano **38** (387), 319-320, 1969; German 1, 918, 223 – filed April 10, 1969.
33. Plumpe, W.H., Jr. U.S. 3, 959, 108 – May 25, 1976, filed December 27, 1971.

6
Fluoride and Complex Fluoride Catalyzed Baths

While the commercial introduction of chromium plating in 1925-1926 was based upon chromic acid baths catalyzed with sulfate only (and these are called conventional baths) newer processes are in large measure related to the use of combinations of sulfate with fluorides or complex fluorides. Fluoride catalysts were recognized as having potentialities at an early stage [1-4], but their actual commercial use has been a long, slow development, depending partly on the increasing availability of new materials of construction of greater corrosion resistance, such as the plasticized polyvinyl chloride used for tank linings beginning in 1937.

Haring and Barrows in 1927 [5] reported that fluoride added as sodium fluoride gave results similar to sulfate, but that it was apparently a weaker catalyst than sulfate, and had to be added in larger quantities, whereas the exact opposite is true. This was because they used a glass container, and in the strongly acid chromic acid medium the sodium fluoride forms hydrofluoric acid which attacks glass rapidly even at low concentrations, and is converted immediately to silicofluoride, a weaker catalyst than sulfate. Actually, fluoride or hydrofluoric acid is the strongest catalyst in use, and is an extremely reactive chemical that will react with many materials to form complex fluorides, which are all weaker catalysts than sulfate. Thus, for example, if nickel solution containing boric acid is dragged into a chromium bath containing fluoride, borofluoride or fluoborate is formed which is such a weak catalyst that it cannot be used alone to make up a satisfactory chromium plating bath, and generally gives rough deposits contaminated with oxide. For this reason, fluoride must be considered as an unstable catalyst in a chromium bath, due to its extreme reactivity.

Muller and his students published three papers [6-8] of a practical nature on the use of fluoride as a catalyst in 1932-1934. Muller and Rossow [6] reported as advantages higher throwing power and a brightening effect on cold chromium plate. Disadvantages included lower current efficiency, a tendency to rough, nodular deposits with thicker plates, and formation of a high resistance film on the lead anodes.

Muller and Vestdal [7] investigated the high resistance film on lead anodes, and on lead-antimony and lead-bismuth alloy anodes, and concluded that good results could be obtained by using lead anodes alloyed with 20 to 30% antimony.

Muller and Drechsel [8] made a more detailed investigation of fluoride-catalyzed baths, but unfortunately used a glass container, and concluded the same as Haring and Barrows that fluoride was a weaker catalyst than sulfate and could be used in larger quantities. Thus, they failed to note any real improvement in current efficiency, but did feel that the fluoride catalyst had a future for better throwing power and wider bright range, even though many of their deposits had scattered tiny warts.

In 1932 Webersinn and Fink made a systematic survey of solutions catalyzed with fluoride, fluosilicate and fluoborate radical, but this was never published [9]. Wax-lined beakers were used, but presumably the solutions were stirred and the temperature taken with glass thermometers, and this would decrease the effectiveness of the simple fluoride radical. Nevertheless, the fluoride radical was found to be stronger than sulfate, and silicofluoride weaker. Fluoborate did not give good deposits except in more dilute solutions containing 150 g/l chromic acid. Current efficiencies were noted to be a little higher than in chromic acid sulfate baths, especially with silicofluoride catalyst, but deposits often tended to be "pimpled."

In 1933 Perlenfein reported good results and slightly higher efficiency with baths with 8 to 10 grams per liter of sodium fluoride [10]. He used waxed containers but glass thermometers, and failed to correlate the sodium fluoride concentration with the chromic acid concentration.

Assmann [11] obtained about one third lower current efficiency, but bright deposits in tests with cold chromium plating with fluoride-catalyzed baths. Curschmann and Heinrich [12] obtained best results with a 250 g/l chromic acid solution containing 8 g/l sodium fluoride in a glass jar at 45°C (113°F). The bright plating range was considerably wider than that of a sulfate-catalyzed bath, and the current efficiency gradually rose to 19-20% compared to about 13% in the sulfate bath.

It should be noted that much of the activity with fluoride-catalyzed baths in the 1930s and even somewhat later was for the purpose of avoiding infringement of the Liebreich patent in Europe or the Fink patent in America. This to some extent determined the nature of the investigations conducted.

Thus, the Panzer chromium plating bath was introduced in Germany early in 1933 by the firm Fr. Blasberg [13]. This was a cold chromium plating bath claimed to give bright deposits. It probably had some connection with Muller and Rossow's work published a few months earlier [6]. According to a German patent application [14] the bath contained 300 g/l chromic acid, 2.5 g/l sodium antimony fluoride and 0.3 g/l of hydrofluoric acid. More than 0.3 g/l HF was considered to tend to nodule formation and low current efficiency, but the 0.3 g/l HF could be replaced with an equal quantity of sulfuric acid without any considerable change in results. While this bath was introduced commercially, it would seem to have only been used for a few years. The plate from cold chromium baths is generally lacking in brightness and would require some buffing; there is nearly always a lack of adhesion also.

During the 1930s and even through the 1940s a typical decorative chromium plating solution based upon silicofluoride additions would have the following composition:

CrO_3	250 g/l
$SO_4^{=}$	0.5 g/l
$SiF_6^{=}$	4.0 g/l

The sulfate was kept as low as possible so that the bath would be as different as possible from the patented sulfate-catalyzed solutions. Good results were obtained in decorative plating, but such baths could not be used for hard chromium plating due to the roughness of thick deposits, which commonly took the form of scattered pimples or black nodules. Later on (1948) the writer found that these scattered pimples could be avoided by the use of dilute solutions with less than 200 g/l CrO_3, but this was unknown during the period in question.

The most detailed report of results with solutions highly catalyzed or over catalyzed with HF and H_2SiF_6 was that of J. Fischer published in 1940 [15]. Fischer stated a common conclusion of the time, that it appeared as if the scattered pimples were caused by the deposition of SiO_2 particles, either from glass dissolved by HF or else from decomposition of H_2SiF_6 into H_2SiO_3. Some tests which he made with containers lined with polyvinyl chloride plastic indicated that this was true, but still the effect was found impossible to avoid in practice.

Fischer was granted two German patents [16, 17] in which he attempted to avoid the scattered pimples or nodules by having some excess free HF present, but this presumably did not work out in practice. The first patent [16] specified a ratio of 1.9 to 4.5 parts of HF to 1 part of SiO_2, and 1 to 4 parts of HF to 1 part of H_2SiF_6. The second patent [17] covered 1 to 4 parts of HF to 1 part of fluoboric acid in chromium baths. However, no complete bath formulations were given.

Perhaps the first to combine the good qualities of sulfate and silicofluoride catalyzed baths with a certain degree of success for hard chromium plating was Karl Gebauer, who described the Fr. Blasberg FBM Hard Chromium Plating Bath in Germany in

January, 1941 [18]. The current efficiency curve of this bath compared to a plain sulfate-catalyzed bath with an optimum composition of about 91 ratio, is given in Fig. 17. The composition of the bath was not published and it was never patented so far as is known. However, the composition may be presumed to be close to the composition used later by Wahl and Gebauer [19] for the preparation of specimens for hardness and wear testing, which was as follows:

CrO_3	250 g/l
H_2SO_4	1.5 g/l
H_2SiF_6	2.5 g/l

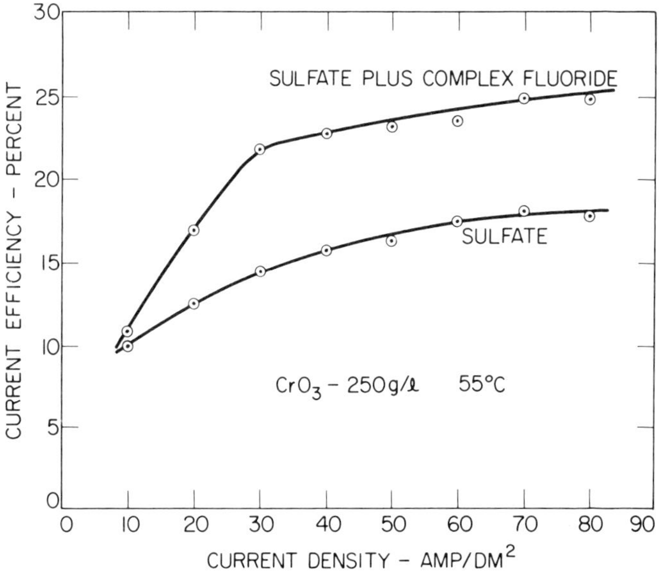

Fig. 17. Gebauer's current efficiency curves for sulfate baths and baths catalyzed with a mixture of sulfate and complex fluoride.

Bilfinger [20, 21, 22] confirmed the value of this formulation, but increased the H_2SiF_6 content to about 4.0 g/l. While he developed methods for determing the hydrofluoric acid and hydrofluosilicic acid content of the baths containing these catalysts, he reported that such baths decomposed more or less rapidly according to the rate of working expressed in amperes per liter. The bath then had to be either partially or wholly discarded and renewed. Such decomposition of baths containing fluoride or silicofluoride had previously been reported in tests at 25°C by Biryukov and Timokhina [23], who stated that the baths gave dull and defective deposits even though the analysis was the same, and the effect could not be overcome by filtration or reoxidation of trivalent chromium.

The reason for the "decomposition" of silicofluoride containing baths is to be found in the use of tanks lined with wire glass sheets to prevent short circuits between the electrodes, as was common at this time. The practice is described in detail by Pfanhauser in 1941 [24]. American practice was similar, and tanks lined with a layer of acid proof brick were also widely used.

The widespread reports of "decomposition" of silicofluoride containing solutions were deprecated strongly by Gebauer and Sommer in 1947, and several life tests were reported to show that no such decomposition occurred [25]. However, the tank linings used were not reported. The reports of the decomposition of silicofluoride containing baths in practice were attributed to excessive iron contamination.

The writer's experience with these matters began

in 1934. Extensive tests were made to determine the best composition of solution for the barrel chromium plating of small parts in bulk. It was found that solutions containing some fluoride or complex fluoride catalyst were invariably better for this purpose than solutions with the sulfate catalyst alone [26]. The difference was relatively small and appeared to be associated with the fact that fluoride containing solutions did not leave a brown chromate film on unplated areas. Thus, when these areas were later exposed to the plating action due to tumbling of the parts in the rotating barrel, they were clean and bright and acquired a good bright chromium plate. In sulfate-catalyzed baths, on the other hand, the areas not plated with bright chromium promptly became covered with irregular patches of brown, iridescent oxide film which spoiled the brightness of chromium applied to these same areas later in the cycle. The result was always dull and poorer appearing work from sulfate than from fluoride catalyzed solutions.

The barrel chromium plating solutions were adjusted to a somewhat higher ratio of chromic acid to catalyst than ordinary still tank solutions, generally about 110 to 120 to 1, and this and the nature of the operation made them distinctly more sensitive to small changes in the catalyst concentration. Thus the barrel chromium baths had to be maintained within a narrower range of composition, or the capacity of the chromium plating barrel would be decreased radically. In June, 1935 it was found that a barrel chromium plating solution made up with commercial magnesium silicofluoride as the source of the silicofluoride catalyst appeared somewhat "oversulfated" compared to an identical solution made up with 30%

c.p. fluosilic acid as the source of the catalyst. It appeared as if the commerical magnesium silicofluoride contained some plain fluoride which is a stronger catalyst. This was confirmed when similar results were obtained from the two solutions after hanging a glass plate in the solution made up with commerical magnesium silicofluoride, for a few hours, after which it was found that both solutions gave the same results. It appeared that the magnesium silicofluoride did give a slightly stronger catalytic effect than c.p. fluosilicic acid, which could readily be neutralized or counterbalanced by the introduction of a small amount of silica.

In January, 1937 the writer determined the approximate sulfate equivalent of fluoride additions in barrel chromium plating solutions. It appeared that one gram per liter of fluoride had about the same catalytic effect as four grams per liter of sulfate. Earlier tests had indicated that silicofluoride was about an eight times weaker catalyst, and that one gram per liter of silicofluoride had about the same catalytic effect as only about one-half gram per liter of sulfate.

These experiences led to the theory that possibly the catalytic effect of silicofluoride or other complex fluorides was due to the amount of hydrofluoric acid or fluoride ion which they introduced into the strongly acid chromic acid plating baths by hydrolysis, as exemplified by the following equation:

$$Na_2SiF_6 + H_2Cr_2O_7 + 2H_2O \rightleftarrows Na_2Cr_2O_7 + SiO_2 + 6HF$$

or more simply

$$H_2SiF_6 + 2H_2O \rightleftarrows SiO_2 + 6HF$$

Liebreich expressed the thought in 1934 [27] that possibly the lower catalytic effect of fluosilicic acid than hydrofluoric acid was due to fewer free fluorine ions coming from fluosilicic acid. Later, in 1940, Fischer expressed the same idea [15].

There would seem to be more than a grain of truth to this idea. In order to investigate it more quantitatively, Frank Passal made some tests in May, 1937, in which he determined the loss in weight of pieces of soft glass tubing 1.5″ long and 5/16″ diameter in 500 ml of chromium plating solution at room temperature for 24 hours. The solution contained 400 g/l CrO_3, 0.4 g/l SO_4, and 6 g/l $SiF_6^{=}$ added as 30% c.p. H_2SiF_6. The loss in weight was compared to the loss in weight with fresh quantities of the same solution with known additions of F^- as NaF, of 0.1 g/l in one instance and 0.4 g/l in another. The results are plotted in Fig. 18, and extrapolation of the plot indicates about 0.5 g/l of fluoride ion already in the solution, from hydrolysis of the H_2SiF_6 in the original make up, into HF.

These are rather crude measurements, with a rather long extrapolation. A calculation based upon the above result indicates that about 10.4% of the fluoride in the c.p. 30% H_2SiF_6 used hydrolyzed into hydrofluoric acid in the chromium plating solution. On the basis of this and other measurements it was concluded in April, 1938 [28] that about 10 to 15% of the H_2SiF_6 in a chromium plating bath is normally dissociated into HF. This is in quite good agreement with the figure of fluoride being about 8 times as strong a catalyst as silicofluoride, corresponding to about 12.5% dissociation or hydrolysis.

Thus, it seems a reasonable assumption, more or less well proven, that the catalytic effect of silicofluoride is due to the amount of fluoride or hydrofluoric

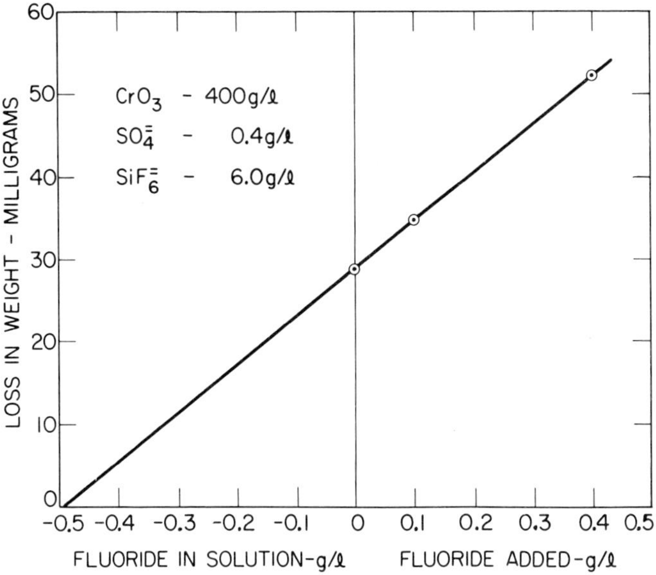

Fig. 18. Passal's glass etching tests to determine the concentration of free HF in a chromium plating bath containing about 6 g/l H_2SiF_6.

acid which it furnishes. This is also probably true of other complex fluorides.

The behavior was demonstrated in good detail in the same work with Frank Passal [28, 29] by taking portions of a chromium plating solution overcatalyzed by the addition of about 13.5 g/l of $SiF_6^{=}$ as 30% c.p. H_2SiF_6, and adding graduated amounts of a pure sodium silicate or water glass solution to neutralize increasing amounts of hydrofluoric acid, or to suppress the amount of it available in the solutions. The solutions were made up to contain about 290 g/l CrO_3 with c.p. chromic acid, and each contained about 0.54 g/l of sulfate. Each solution was analyzed for total silicon and total fluorine content with the results shown in Table 3.

Table 3

	ml/l of Na_2SiO_3 soln. containing 14.3 g/l Si	SiF_6 equivalent of Na_2SiO_3 added – g/l	SiF_6(Si) g/l	SiF_6(F) g/l
1	none	none	13.1	13.5
2	7.5	0.542	14.0	13.5
3	15.0	1.085	14.4	13.5
4	22.5	1.627	14.6	12.9
5	30.0	2.170	15.4	13.3

The results of current efficiency tests on these five solutions at 35°C with nickel cathodes are plotted in the curves shown in Fig. 19. For comparison, the current efficiency-current density curves of similar solutions at 100/1 and 50/1 chromic acid to sulfate ratio are shown with dotted lines. It can be noted that the solution No. 1 with no Na_2SiO_3 added behaved similarly to the 50/1 ratio solution containing sulfate alone, perhaps showing about a 40/1 ratio behavior. As increasing amounts of Na_2SiO_3 were added the current efficiency increased with each addition until with solution No. 5 the current efficiency was almost up to that of a 100/1 solution containing only sulfate.

Three additional tests were made [28] with solutions with 45, 69 and 75 c.c. of the Na_2SiO_3 solution added, but the current efficiencies were almost identical with solution No. 5, within the experimental error. The bright plate range decreased with excessive additions of Na_2SiO_3, and with 60 and 75 c.c. of

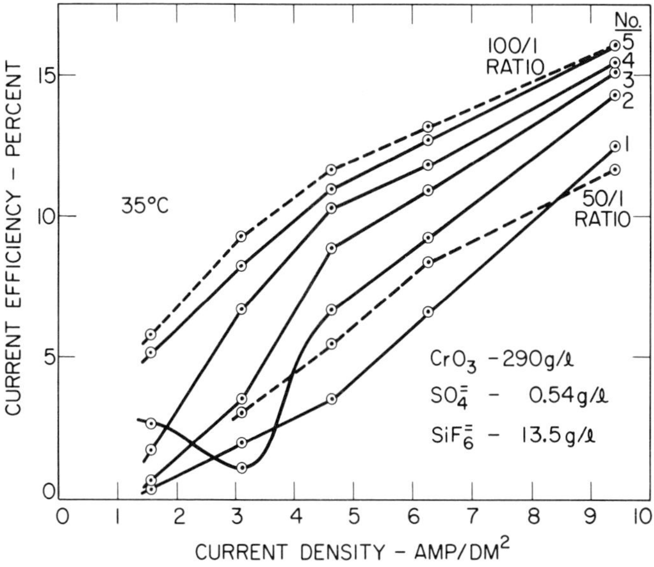

Fig. 19. Dubpernell's current efficiency determinations showing the gradual conversion of an over-catalyzed silicofluoride-containing bath (No. 1) with about 40:1 ratio behavior, to a behavior similar to 100:1 ratio (No. 5) with successive small additions of Na_2SiO_3 solution.

Na_2SiO_3 solution added a small amount of silica gel could be seen floating about, indicating saturation. There was a very small decrease in current efficiency with the final addition of 75 c.c. of Na_2SiO_3 solution.

These measurements show that the catalytic effect of silicofluoride to permit production of chromium plate from chromic acid solution is controlled by the amount of free hydrofluoric acid which it introduces, and that this in turn can be substantially modified by rather small amounts of silicate added to the solution. Thus, solution No. 1 in Fig. 19 exhibited about a 40/1 ratio behavior, but this was converted to about a 90/1 behavior by the addition of 0.43 g/l Si or about 1.87 g/l Na_2SiO_3, a non-catalytic material.

Taking the sulfate content of these solutions as 0.5 g/l and calculating the relative strength of the silicofluoride in each instance, we find that it was about 1/2 as strong as sulfate in solution No. 1 as made up, but only about 1/5 as strong as sulfate in solution No. 5 after the addition of 1.87 g/l of sodium silicate. This checks well with the calculations made [29] of the average strength of the $SiF_6^{=}$ catalyst in three commercial chromium plating baths in brick lined tanks, which came out about 1/4.6 compared to sulfate. Another check from German experience is given by Brugger [30] who mentions that lg. of sulfate is equivalent to 5 g. of silicofluoride.

Thus, it is apparent that the catalytic effect of silicofluoride varies on a sliding scale from 1/2 that of sulfate in new baths kept in plastic lined tanks to about 1/5 that of sulfate if the solution is in a glass or brick lined tank with which the true catalyst, the HF formed, may react freely and pick up excess silica. Actually, there is a continuous scale of catalytic effect extending from additions of pure HF, which is about four times as strong as sulfate, all the way to silicofluoride containing excess silica, which has only about 1/5 the effect of an equal weight of sulfate. If HF is free to dissolve silica it does so rather rapidly, and is converted to fluosilicic acid, and finally to fluosilicic acid containing excess silica.

Thus, hydrofluoric acid is about 20 times as strong a catalyst as fluosilicic acid saturated with silica or silicon. This is diagrammed in Fig. 20 using

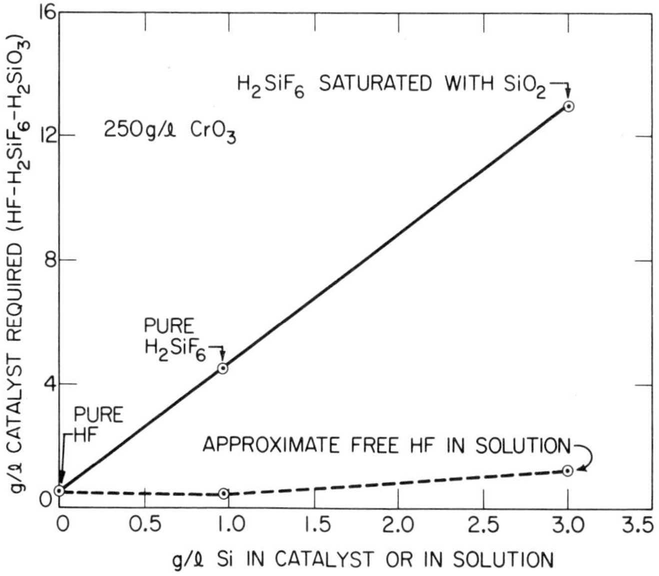

Fig. 20. Approximate effect of increasing amounts of silicon on the catalyst strength of fluoride catalysts.

the three points which have been determined more or less quantitatively. The approximate amount of free HF in solution is also included in Fig. 20, based upon 10% dissociation or hydrolysis of H_2SiF_6 into HF, and it can be noted that this is a relatively constant quantity. Closer investigation might reveal that it is really constant for constant catalytic effect. The ability of fluosilicic acid to dissolve some excess silica had been noted as far back as 1908 by A.G. Betts in his book "Lead Refining by Electrolysis," page 35.

High-silica fluosilicic acids were investigated in some detail by S.M. Thomsen [33]. Thomsen found that fluosilicic acid could contain up to 18% more silica than represented by the formula H_2SiF_6. This is in good agreement with the measurements reported in Fig. 19.

The importance of this effect has been questioned by many workers throughout almost the entire history of chromium plating, because any etching or pitting of glass or bricks by a chromium plating solution is rarely visible, and is discounted for this reason. However, this point of view neglects the fact that the catalyst in a chromium plating bath is present in very small concentration, even less than 1% by weight of the chromic acid concentration, and any reaction whatsoever affects the strength of the catalyst.

A recent example of the importance of not using glass containers is found in articles by Solov'eva [31] and Solov'eva and Lapshina [32]. Solov'eva [31] noted the effect of contact with glass in several footnotes, but did not experiment with other than glass containers, and reports fluoride to be a weaker catalyst than sulfate.

It is noteworthy that the highest current efficiencies attained in the tests in Fig. 19 do not exceed those obtained with sulfate catalyst alone at 100/1 ratio. In order to obtain the higher current efficiencies shown in Fig. 17, the silicofluoride needs to be free of excess silica, and the sulfate content higher than 0.5 g/l, particularly when the chromic acid content is 200 g/l or higher.

A bright, high speed solution was developed in 1948 which contained 150 g/l CrO_3, 0.5 g/l $SO_4^{=}$ and 3.3 g/l $SiF_6^{=}$. At 131-140°F. (55-60°C) and 3-5 amp/in^2 (46.6 - 77.7 amp/dm^2) excellent results were obtained, but this solution was supplanted by self-regulating baths shortly thereafter.

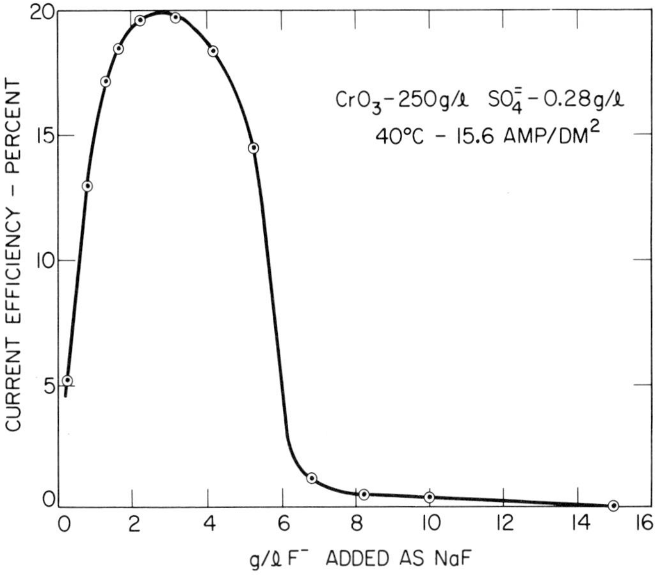

Fig. 21. A Webersinn and Fink current efficiency curve for a solution with increasing amounts of fluoride catalyst only.

The peculiar shape of Curve No. 2 in Fig. 19 was found to be reproducible and not an artifact. The higher current efficiency at the lowest current density was checked in three ways: (1) by the gain in weight, (2) by the loss in weight on stripping, and (3) by titration of the 30 mg of chromium in the strip solution. This peculiar high current efficiency, low current density deposit was rather dull, but appeared to be almost 100% chromium the same as other plates. It probably bears a relationship to the low current density micro-cracking discovered by Chessin [34]. There could also be a connection with the low current density behavior of chloride catalyst discussed in Chapter 4.

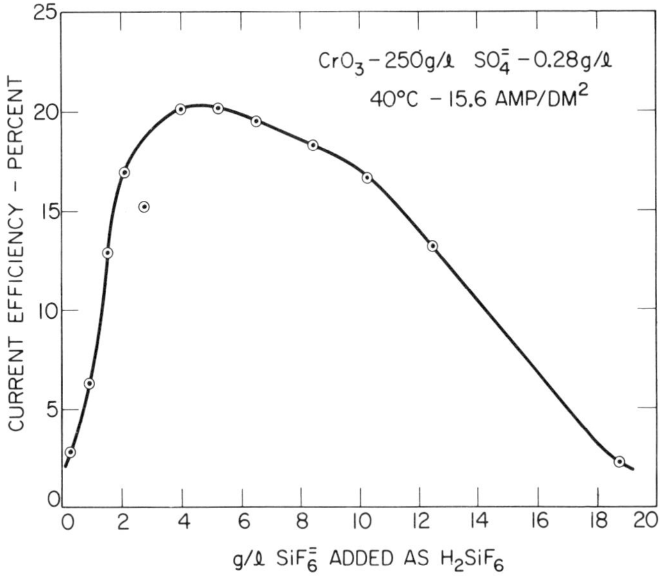

Fig. 22. A Webersinn and Fink current efficiency curve for a solution with increasing amounts of silicofluoride catalyst only.

A number of other complex fluorides are quite similar to silicofluoride. Typical current efficiency-catalyst concentration curves taken from the work of Webersinn and Fink [9] are given in Fig. 21 and Fig. 22 for fluoride and for silicofluoride ion. Passal patented fluoaluminate, $AlF_6^=$ fluotitanate, $TiF_6^=$ and fluozirconate, $ZrF_6^=$ ions as chromium plating catalysts [35]. Some of his measurements with these ions are shown in Fig. 23, 24 and 25.

These other complex fluorides are subject to the same shifting of catalyst strength with the presence of excess of the complexing agent, as is silicofluoride with the addition of excess silica. Thus, if a chro-

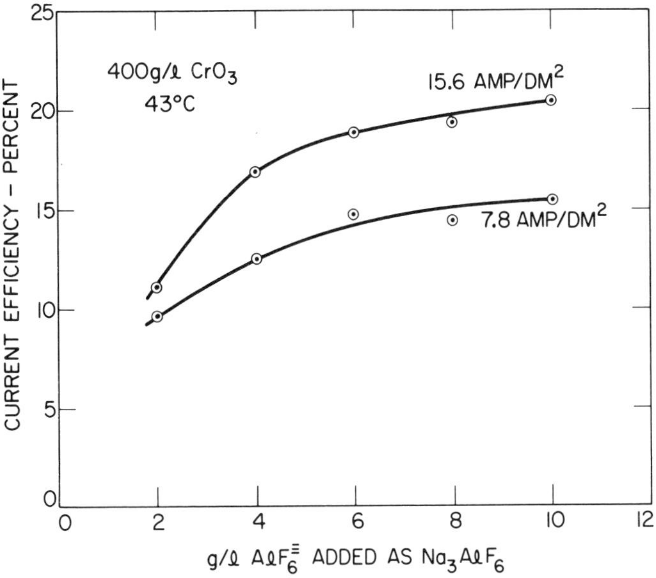

Fig. 23. Passal's current efficiency determinations in a solution with increasing amounts of fluoaluminate catalyst only.

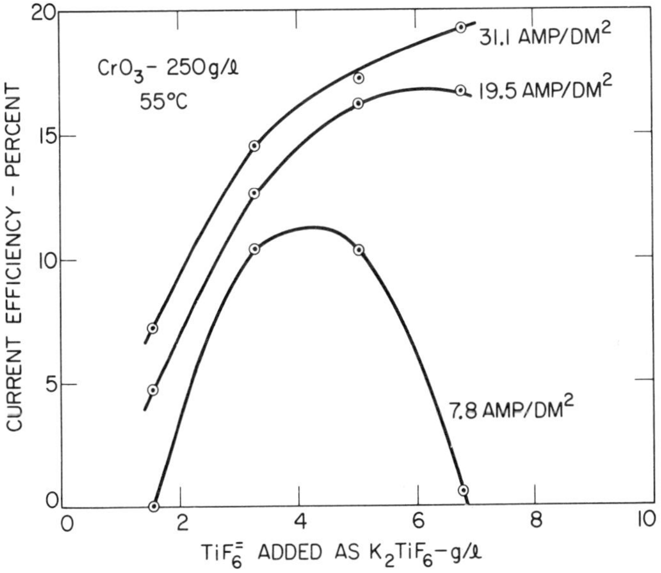

Fig. 24. Passal's current efficiency determinations in a solution with increasing amounts of fluotitanate catalyst only.

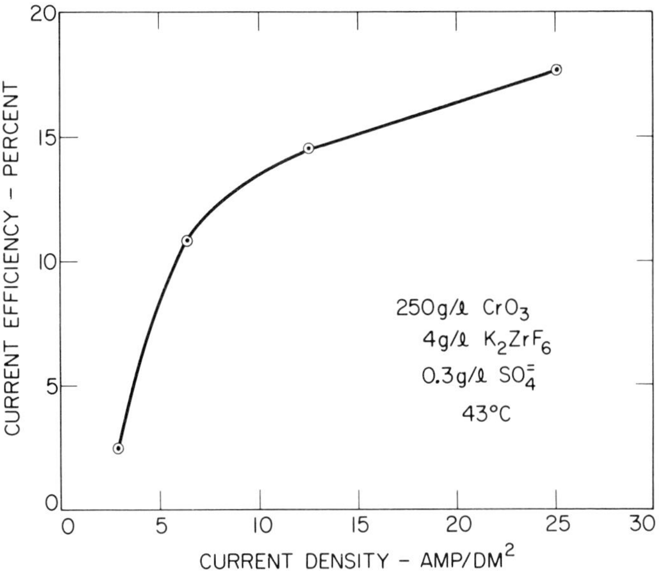

Fig. 25. Passal's current efficiency determinations in a solution with fluozirconate catalyst only.

mium plating bath with fluoaluminate, $AlF_6^{\equiv}$ becomes contaminated with aluminum, the catalyst strength and current efficiency decrease radically.

This is a problem with all baths with fluoride or complex fluoride catalyst. The effective catalyst content tends to shift radically with bath contamination, and proper control and maintenance become extremely difficult or impossible under many conditions. Even dissolved iron, which commonly has substantially no effect in sulfate-catalyzed baths, shifts the equilibrium in fluoride-catalyzed solutions.

The effect of boric acid dragged in from nickel solutions has already been mentioned. Boric acid

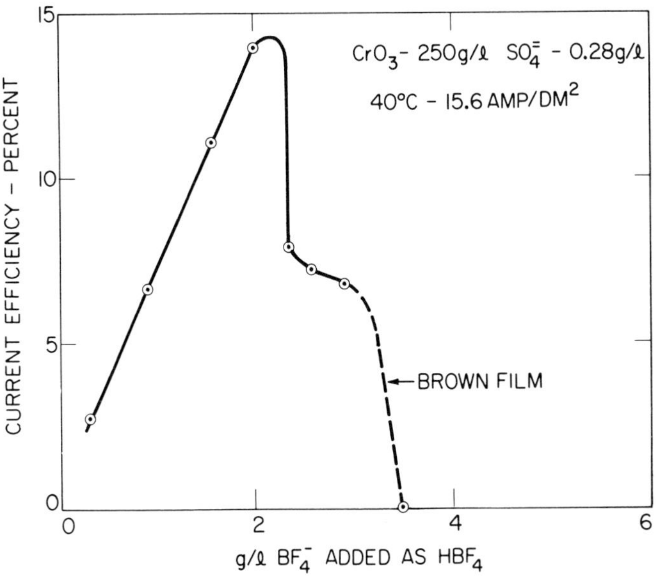

Fig. 26. A Webersinn and Fink current efficiency curve for a solution with increasing amounts of borofluoride (fluoborate) catalyst only.

forms such a stable complex with fluoride that satisfactory chromium plating can no longer be obtained unless other catalysts are also present. This behavior of borofluoride or fluoborate is evident from Fig. 26 taken from the work of Webersinn and Fink [9]. The deposits from such solutions are generally rough and dull, and contaminated with oxide.

The use of these other fluoride-containing anions as alternative to silicofluoride has become more common. Thus, Stareck, Seyb, Johnson and Rowan [36] give exact details for the production of bright crack-free chromium plate using fluoaluminate, fluotitanate, or fluozirconate ion instead of silicofluoride. Chessin [37] uses fluoborate, fluoaluminate, fluophosphate, fluozirconate and fluotitanate in plating microcracked chromium. Seyb and co-workers mention these alternative complex fluorides in plating microcrack chromium [38] and in plating rapidly moving steel strip [39]. Roggendorf includes them in a patent on cold chromium plating [40], and in a later patent uses fluoantimonate and fluostannate for the same purpose [41]. Law and Jones [42] also use such catalysts for plating microcracked chromium. Becking [43] describes the use of rare earth phosphofluorides.

Ivanova and Zosimovich [44] describe the structure of chromium deposits made over a wide range of temperatures and current densities in baths with fluoride, fluotitanate, fluozirconate and fluoniobate ions, and conclude correctly that the effective catalyst appears to be the amount of free fluoride ion and not the particular complex ion supplying it. While Zosimovich and Ivanova [45] mentioned earlier that they used polyethylene containers, these same authors later reported [46] typical difficulty with the "aging" of their solutions and the decrease of current efficiencies with time. Used baths containing as much as 5.25 g/l total fluoride were found to resemble fresh baths containing only 0.8 g/l F^-. This was incorrectly ascribed to the formation with time of electrochemically inactive chromium-fluorine complexes, rather than any of the many other complexes possible, such as those studied in their later article [44].

The instability or "aging" of chromium plating baths catalyzed with fluorides or complex fluorides is

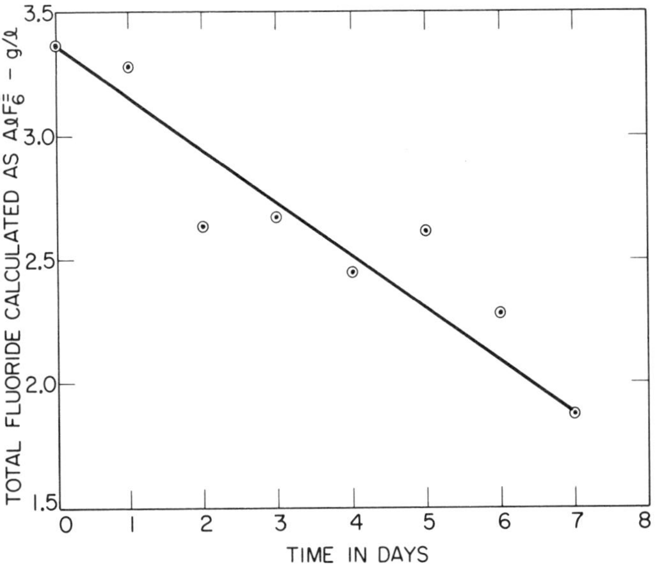

Fig. 27. Loss of fluoride during electrolysis of a three liter chromium plating bath containing 320 g/l CrO_3 and 4.8 g/l $(NH_4)_3AlF_6$, at 150°F, using 34 amperes.

still a common problem, since the active catalyst is the fluoride ion, or the hydrofluoric acid which this forms in the chromic acid solution, which in turn is difficult to maintain in a free condition. Even the slag particles in the steel basis metal or the silicon in cast iron may give some access to silica and cause a slow conversion to silicofluoride.

This catalyst (fluoride) may also be considered unstable or be lost in small quantities from time to time due to its volatility. Following the publication of the method of Willard and Winter in 1933 [47] for the determination of total fluoride by distillation as

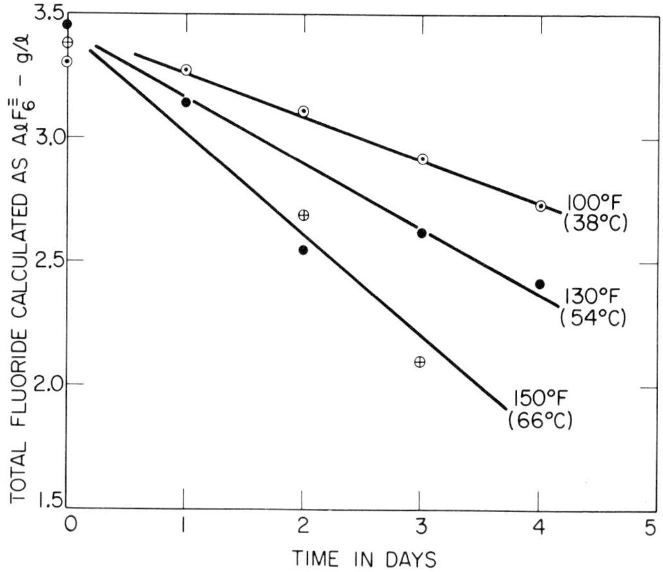

Fig. 28. Loss of fluoride upon bubbling air through two liters of solution containing 320 g/l CrO_3 and 4.8 g/l $(NH_4)_3AlF_6$ at different temperatures.

H_2SiF_6, it was thought that there could be losses due to volatilization of this acid if the solution was used too hot for too long a period. However, the work of Munter, Aepli and Kossatz in 1947 [48] and of Fox and Jackson in 1959 [49], showed that the volatile constituent was hydrofluoric acid itself.

This was confirmed by life tests made by Tulumello and Passal on fluoaluminate catalyst compared to baths containing fluosilicic acid, which showed a continuing decrease with time of the total fluorine content in all baths [50]. No loss due to volatility either of silicon or aluminum was noted in any of the tests. Tulumello's results for the loss of fluoride during electrolysis are shown in Fig. 27 and the loss due to merely bubbling air through the solution at various temperatures in Fig. 28. This loss of hydrofluoric acid due to its volatility is a strong reason for not using air agitation in baths with fluoride or complex fluoride catalyst.

The older work on the chemistry of fluorides and complex fluorides was reviewed by Kausch in 1936 [51]. A more up to date discussion of this type of chemical compound is found in Simons' two volume work [52].

One generality which is worth keeping in mind is that practically all fluoride compounds are more soluble, and may be several times more soluble in chromium plating solution than in pure water. Calcium fluoride is many times as soluble. Thus, the textbook values for solubility in water are only a very rough guide, and the actual solubility in chromic acid plating solution has to be determined separately in each case. It would be worthwhile to have a complete table of such solubilities.

References

1. Hambuechen, C., assignor to Electrometallurgical Co., U.S. 1, 544, 451 – June 30, 1925.
2. Fink, C.G., and McLeese, H.D., assignors to United Chromium, Incorporated, U.S. 1, 844, 751 – February 9, 1932, and U.S. 1, 928, 284 – September 26, 1933.
3. Sohn, E., and Paddock, R.J., assignors to Standard Sanitary Mfg. Co., U.S. 1,815,081 – July 21, 1932.
4. Ewing, D.T., U.S. 1, 864, 014 – June 21, 1932.
5. Haring, H.E. and Barrows, W.P., "Electrodeposition of Chromium From Chromic Acid Baths," Technologic Papers of the Bureau of Standards, No. 346, Volume 21, 413-449, 1927.
6. Muller, E. and Rossow, O., Zeit. Elektrochem. **38**, (11), 883-884, 1932.
7. Muller, E., and Vestdal, J.E., Zeit. Elektrochem. **40**, (1) 14-18, 1934.
8. Muller, E., and Drechsel, H., Zeit. Elektrochem. **40**, (110), 707-713, 1934.

9. Webersinn, T.H., and Fink, C.G. "Electrodeposition of Chromium From Chromic Acid-Fluorine Solutions," United Chromium Incorporated Report No. 123, 65 pages, February 20, 1932.
10. Perlenfein, A. "A Chromium Plating Bath With the Fluoride Ion," Rensselaer Polytechnic Institute Bulletin No. 39, 36 pages, Troy, New York, February, 1933.
11. Assmann, K., Chem. Ztg. **59**, (17), 177-178, February 27, 1935.
12. Curschmann, E., and Heinrich, H., Monthly Review Am. Electroplaters' Soc, **23**, 52-55, February, 1936.
13. Fr. Blasberg advertisement, Metallwaren Industrie und Galvano - Technik **31**, pages 41, 47-48, February 1, 1933.
14. Ly Blasberg, German patent application B158,681 – December 8, 1932; opened to public inspection, August 9, 1934.
15. Fischer, J., Wiss. Veroffent. Siemens-Werken 1940, pgs. 138-168, "The Effect of Hydrofluoric Acid and Fluosilicic Acid On The Electrodeposition of Chromium From Solutions of Chromic Acid, As Compared To Sulfuric Acid."
16. Fischer, J., assignor to Siemens & Halske, A.G., German patent 710, 694, filed July 5, 1939; published September 19, 1941.
17. Fischer, J., (assignor to Siemens & Halske A.G.), German patent 746, 236 filed January 17, 1941; published June 8, 1944.
18. Gebauer, K., Oberflaechentechnik **18**, 2-3, 11-13, 19-22, 31-33, 1941; abstract: Metallwirtschaft **20** (45), 1102-1103, 1941; abstract Translation No. 1431, 7 pages, Henry Brutcher, Altadena, California.

19. Wahl, H., and Gebauer, K., Metalloberflaeche **2**, 25-37, 1948; Translation No. 2700, 18 pages, Henry Brutcher, Altadena, California.
20. Bilfinger, R., "Das Hartverchromungs Verfahren" (The Hard Chromium Plating Process), 2nd edition, 235 pages, Leipzig: Herm. Beyer Verlag, 1942.
21. Bilfinger, R., Metallwirtschaft **22**, (33-35) 466-472, 1943; translation No. 1704, Henry Brutcher, Altadena, California.
22. Bilfinger, R., Metallwaren – Industrie u. Galvanotechnik **42** (5), 129-131, 1944 (reprint of part of reference 21).
23. Biryukov, N.D., and Timokhina, A.V., Korroziya i Borba s Nei **6**, (2), 45, 1940; C.A.**36**, 3733.
24. Pfanhauser, W., "Galvanotechnik" (Electroplating), 8th edition, 1591 pages in two volumes, Leipzig: Akademische Verlagsgesellschaft, 1941.
25. Gebauer, K., and Sommer, K., Metalloberflaeche **1** (2), 25-27, 1947; Translation No. 1958, 12 pages, Henry Brutcher, Altadena, California.
26. Dubpernell, G., and Martin, S.M., Proc. Amer. Electroplaters' Soc. **34**, 180-200, 1947.
27. Liebreich, E., Zeit. fur Elektrochemie **40**, 73-87, 1934.
28. Dubpernell, G., "Chromic Acid – Fluoride-Silicofluoride-Sulfate Chromium Plating Baths", 17 pages, special report, United Chromium, Incorporated, Waterbury, Connecticut, April 1938.
29. Dubpernell, G., "Great Lakes Plating and Japanning Company-Evidence," 19 pages, special report, United Chromium, Incorporated, Waterbury, Connecticut, April 13, 1938.

30. Brugger, R. (to Deinert & Co.), German 1,078,840 – filed December 24, 1956; published March 31, 1960.
31. Solov'eva, Z.A., Elektrokhimiya **1** (7), 812-817, 1965; Soviet Electrochemistry **1** (7), 721-725, 1965.
32. Solov'eva, Z.A. and Lapshina, A.E., Elektrokhimiya **1** (8), 941-946, 1965; Soviet Electrochemistry **1** (8), 840-844, 1965.
33. Thomsen, S.M., J. Am. Chem. Soc. **72**, 2798, 1950; Anal. Chem. **23**, 973-975, 1951; J.Am. Chem. Soc. 74, 1690-1693, 1952.
34. Chessin, H., "Method of Electroplating Microcrack Chromium," U.S. 3,340,165 – September 5, 1967.
35. Passal, F., assignor to Metal & Thermit Corporation, U.S. 2,962,428 – November 29, 1960
36. Stareck, J.E., Seyb, E.J., Johnson, A.A., and Rowan, W.H., assignors to Metal & Thermit Corporation, U.S. 2,952,590 – September 13, 1960.
37. Chessin. H., assignor to M & T Chemicals Inc., U.S. 3,340,165 – September 5, 1967.
38. Mahlstedt, H., Dunn, J.W., Couch, R.W., and Seyb, E. J., assignors to M & T Chemicals Inc., U.S. 3,461,048 – August 12, 1969.
39. Seyb, E.J., Woehrle, R.E., and Neitzel, J.G., assignors to M & T Chemicals Inc., U.S. 3,498,892 – March 3, 1970.
40. Roggendorf, W., U.S. 3,418,220 – December 24, 1968.
41. Roggendorf, W., U.S. 3,661,733 – May 9, 1972.
42. Law, M.J., and Jones H., assignors to Permalite Chemicals Limited, U.S. 3,713,999 – January 30, 1973.

43. Becking, D.H., assignor to Udylite Corporation, U.S. 3,694,328 – September 26, 1972.
44. Ivanova, N.D., and Zosimovich, D.P., Elektrokhimiya **7** (12), 1773-1778, 1971; Soviet Electrochemistry 7 (12), 1714-1718, 1971.
45. Zosimovich, D.P., and Ivanova, N.D., Ukr. Khim. Zh. **33** (2), 200-201, 1967; Bull, Doc. Centre D'Inf. Chrome Dur, April, 1969, pages 19-21.
46. Ivanova, N.D., and Zosimovich, D.P., Ukr. Khim Zh. **35** (10), 1105-1107, 1969; through Chemical Abstracts 72: 27607b.
47. Willard, H.H., and Winter, O.B., Ind. Eng. Chem. Anal. Ed. **5** (1), 7-10, 1933.
48. Munter, P.A., Aepli, O.T., and Kossatz, R.A., Ind Eng. Chem. **39** (3), 427-431, 1947.
49. Fox, E.J., and Jackson, W.A., Anal. Chem. **31** (10), 1657-1662, 1959.
50. Tulumello, A., "A Study of Mixed Catalyst Semi-Self-Regulating Systems for Use in Porous Chromium Plating," R & D Technical Report D-104, Metal and Thermit Corporation, Detroit 20, Michigan, 15 pages, October 1, 1957.
51. Kausch, O., "Flussäure, Kieselflussäure Und Deren Metallsalze" (Hydrofluoric Acid, Fluosilicic Acid, and Their Metal Salts), 438 pages, Stuttgart: Verlag Von Ferdinand Enke, 1936.
52. Simons, J.H., "Fluorine Chemistry," Volume 1, 615 pages, 1950; Volume 2, 565 pages, 1954, New York: Academic Press, Inc.

Name Index

Subject Index